SpringerBriefs in Statistics

JSS Research Series in Statistics

The current research of statistics in Japan has expanded in several directions in line with recent trends in academic activities in the area of statistics and statistical sciences over the globe. The core of these research activities in statistics in Japan has been the Japan Statistical Society (JSS). This society, the oldest and largest academic organization for statistics in Japan, was founded in 1931 by a handful of pioneer statisticians and economists and now has a history of about 90 years. Many distinguished scholars have been members, including the influential statistician Hirotugu Akaike, who was a past president of JSS, and the notable mathematician Kiyosi Itô, who was an earlier member of the Institute of Statistical Mathematics (ISM), which has been a closely related organization since the establishment of ISM. The society has two academic journals: the *Japanese Journal of Statistics and Data Science* (JJSD, Springer), which is the successor of the *Journal of the Japan Statistical Society* (JJSS) and the *Journal of the Japan Statistical Society* (Japanese Series). The membership of JSS consists of researchers, teachers, and professional statisticians in many different fields including mathematics, statistics, engineering, medical sciences, government statistics, economics, business, psychology, education, and many other natural, biological, and social sciences. The JSS Series of Statistics aims to publish recent results of current research activities in the areas of statistics and statistical sciences in Japan that otherwise would not be available in English; they are complementary to the two JSS academic journals, both English and Japanese. Because the scope of a research paper in academic journals inevitably has become narrowly focused and condensed in recent years, this series is intended to fill the gap between academic research activities and the form of a single academic paper. The series will be of great interest to a wide audience of researchers, teachers, professional statisticians, and graduate students in many countries who are interested in statistics and statistical sciences, in statistical theory, and in various areas of statistical applications.

Naoto Kunitomo · Seisho Sato

The SIML Filtering Method for Noisy Non-stationary Economic Time Series

 Springer

Naoto Kunitomo
The Institute of Statistical Mathematics
(ISM)
Tachikawa, Tokyo, Japan

Seisho Sato
Graduate School of Economics
University of Tokyo
Tokyo, Japan

ISSN 2191-544X   ISSN 2191-5458 (electronic)
SpringerBriefs in Statistics
ISSN 2364-0057   ISSN 2364-0065 (electronic)
JSS Research Series in Statistics
ISBN 978-981-96-0881-2   ISBN 978-981-96-0882-9 (eBook)
https://doi.org/10.1007/978-981-96-0882-9

Preface

Over the last few decades a number of statistical methods for the analysis of economic time series data have been developed in econometrics and statistical time series analysis. Nevertheless, there has been no single convincing statistical method for the analysis of economic time series such as major macroeconomic series. This may be due to the fact that the economic time series data contain not only the stationary components such as business cycles, seasonal, and irregular noise components, but also the non-stationary components such as trends, breaks, and structural changes. Important macroeconomic data, which are regularly published and reported in the mass media, are usually compiled from various sources including sample surveys conducted by official statistical agencies, whereas the statistical time series analysis often ignores measurement errors. For instance, quarterly GDP (Gross Domestic Product) is observed once every 3 months and our main interest may be business cycles in say 30 years. Another example would be the monthly Consumers Price Index (CPI), which is observed once in a month. Then there are still gaps between the commonly used statistical time series methods and the observed real economic data. One complicating factor is that economists often use the seasonally adjusted data whereas the official agencies often apply the X-12-ARIMA or X-13ARIMA-SEATS program to produce seasonally adjusted data, which use the univariate Reg-ARIMA model and possibly spectral analysis of the time series to remove the seasonality as the default filtering procedure.

In this monograph, we investigate a new filtering technique, called the SIML filtering, to estimate the hidden states of the trend and cycle components of economic time series, which are non-stationary, and to handle multiple time series data when the sample size of the time series is not large. To address some of the challenges in the analysis of economic time series, Kunitomo and Sato (2017), Kunitomo, Sato, and Kurisu (2018), and Kunitomo, Awaya, and Kurisu (2019) have developed the separating information maximum likelihood (SIML) method for estimating the non-stationary errors-in-variables models. They have discussed the asymptotic and finite sample properties of the estimators of unknown parameters in the statistical models. We use their results to solve the filtering problem of hidden random variables and show that they offer a novel approach to analyzing macroeconomic time series.

The methodology presented in this monograph has originally emerged from our joint work on the volatility estimation in high-frequency financial data. We co-authored a monograph in this JSS-Springer Series with Daisuke Kurisu (Kunitomo, Sato, and Kurisu (2018)[1]). This may sound strange to some statisticians because the macroeconomic data, which constitutes the primary focus of this book, and the high-frequency financial data including common stocks, exchange rates, and interest rates in the first book, are quite different. However, there has been a common feature in two different types of economic data in the sense that both data are a mixture of stationary and non-stationary parts, whereas we want to estimate the invariant quantities of time series in some sense. The focus of this book is on the analysis of discrete time statistical models and observations with measurement errors while the previous book focused on the continuous-time statistical models and discrete observations with measurement errors.

This book is a summary of our joint research project on the SIML method for economic time series and applications over the past few years. We hope that it will give many readers a better understanding of the problem of economic data and that it will also be a good starting point for investigating the unsolved related topics in the future. The R-program called x12simldoc92 developed by Seisho Sato for this book is currently available.[2] This book is also, in a sense, a report on the consultancy projects with official statisticians at ERSI (Economic and Social Research Institute, Cabinet Office of Japan) and Statistics Bureau of Japan, who introduced us to the challenges of real seasonal adjustment and numerous inquiries regarding the use of the X-12-ARIMA and X-13ARIMA-SEATS programs developed by the U.S. Census Bureau for official seasonal adjustment.

During the preparation of the manuscripts, we have received several comments from researchers and we appreciate them. In particular, we are grateful to Akihiko Takahashi for a discussion on the filtering problem in the early stage of our project. We also thank Sakurai Tomoaki for providing some figures in Kunitomo, Sakurai, and Sato (2022).[3] Chapter 4 of this book is a revised version of the published paper with Naoki Awaya and Daisuke Kurisu.[4] We thank Naoki Awaya and Daisuke Kurisu for the permission of its use. This research has been supported by a project of the

---

[1] Kunitomo, Sato, and Kurisu (2018), *Separating Information Maximum Likelihood Estimation for High Frequency Financial Data*, Springer. https://link.springer.com/book/10.1007/978-4-431-559 30-6.

[2] http://www.kunitomo-lab.sakura.ne.jp/x12simldoc92(kuni2023-2-2).pdf.

[3] Kunitomo, Sakurai, and Sato (2022), "A Filtering Method of Economic Time Series and a Macro-Index" (in Japanese), Toukei-Kenkyu-Ihou Vol.79, 1–20, Bureau of Statistics, Japan.

[4] Kunitomo, Awaya, and Kurisu (2020), "Comparing Estimation methods of Non-stationary Errors-in-Variables Models," Japanese Journal of Statistics and Data Science, Springer, Vol.3, 73–101, https://doi.org/10.1007/s42081-018-0017-3.

Consortium for Training Experts in Statistical Sciences (TESS) at ISM (Institute of Statistical Mathematics, Tokyo) and JSPS Grant-in-Aid for Scientific Research No. 17H02513 and 22K01428.

Tokyo, Japan   Naoto Kunitomo
August 2024

Contents

Chapter 1
Introduction

Abstract In this book, we present a novel filtering approach for estimating the hidden states of random variables in the multiple noisy non-stationary time series data. Our methodology is particularly suited to the analysis of small sample non-stationary macroeconomic time series. The method is based on the frequency domain application of the separating information maximum likelihood (SIML) method, which was developed by Kunitomo et al. (Separating information maximum likelihood estimation for high-frequency financial data. Springer, 2018) and (Jpn J Stat Data Sci 1:297–332, 2020), and Nishimura et al. (2019). We propose to use the filtering method of hidden random variables of trend-cycle, seasonal, and measurement error components.

Extensive research has been conducted on the use of statistical time series analysis for macroeconomic time series. A salient feature of macroeconomic time series that distinguishes it from standard statistical time series analysis is that the observed time series is an apparent mixture of non-stationary and stationary components. The second feature is that the measurement errors in economic time series play an important role, because macroeconomic data are typically compiled from different sources, including sample surveys in major official statistics, while the statistical time series analysis often ignores measurement errors. Thirdly, the sample size of macroeconomic data is relatively limited, and we have about 120 time series observations for each series for quarterly data over 30 years. The quarterly GDP series, which is the most important data in the Japanese macroeconomy, for instance, is regularly estimated and released once in every three months by the Cabinet Office of Japan.[1] Fourthly, to publish the seasonally adjusted data, the official agencies in Japan usually

---

[1] Currently, GDP time series data in Japan are constructed and seasonally adjusted since 1994Q1, although some may misunderstand the historical GDP data prior to 1994 as if they were constructed and measured exactly in the same way. The measuring procedures of official GDP series have been changed several times including the base-year changes. See https://www.esri.cao.go.jp/index-e.html for the detailed explanation of how GDP data in Japan are constructed.

© The Author(s), under exclusive license to Springer Nature Singapore Pte Ltd. 2025   1
N. Kunitomo and S. Sato, *The SIML Filtering Method for Noisy Non-stationary Economic Time Series*, JSS Research Series in Statistics,
https://doi.org/10.1007/978-981-96-0882-9_1

apply the X-12-ARIMA software,[2] which uses the univariate Reg-ARIMA model to remove the seasonal components as the standard filtering procedure. Given the limited sample size, it is of the utmost importance to employ an appropriate statistical procedure to extract information on trend-cycles, seasonal, and noise (or measurement errors) components in a systematic manner from multiple time series data.

This book presents a novel filtering procedure for estimating the hidden states of trend-cycle, which are non-stationary, and for handling multiple time series data, when we have small sample time series. Originally, Kunitomo and Sato (2017) and Kunitomo et al. (2018, 2020) have developed the separating information maximum likelihood (SIML) method for estimating the non-stationary errors-in-variables models. The authors have discussed the asymptotic and finite sample properties of the estimation method for unknown parameters in the statistical models. We use their results to solve the current filtering problem of hidden random variables and show that they lead to a new way to handle macroeconomic time series.

The literature on the non-stationary economic time series analysis includes the works of Engle and Granger (1987) and Johansen (1995). These studies examined multivariate non-stationary and stationary time series and developed the notion of co-integration without measurement errors. The focus of their analysis is on the time domain of time series. Our work is related to that of the aforementioned authors, but it has different aspects. Our focus is on the non-stationary trend-cycle, seasonal, and measurement error components in the non-stationary errors-in-variable models and their frequency domain analysis. Some related econometric studies on time series in the frequency domain are Baxter and King (1999), Christiano and Fitzgerald (2003), and Müller and Watson (2018).

Another related issue is the statistical analysis of measurement errors. In the field of statistical multivariate analysis, notable studies on the errors-in-variables models include those by Anderson (1984, 2003) and Fuller (1987). However, these studies focused on multivariate cases of independent observations, which differ from the underlying situation presented in this study.

In the context of statistical filtering methods, Kitagawa (2021) discussed the standard statistical methods already known, including the Kalman-filtering and particle-filtering methods. Although many studies have examined statistical filtering theories, we must exercise caution in analyzing non-stationary multivariate economic time series. See Granger and Hatanaka (1964), Brillinger and Hatanaka (1969) on early studies, and Harvey and Trimbur (2008) on the relationship between HP filter and other methods, for instance. Here we should mention two issues. First, the existing methods often depend on the underlying distributions such as the Gaussian distributions for the Kalman-filtering, and second, they often depend on the dimension of state variables. There may be some difficulty in extending the existing methods to high-dimension cases, even when the dimension is about 10. On the other hand,

---

[2] The latest version of seasonal adjustment program developed by the U.S. Census Bureau is X-13ARIMA-SEATS, yet there is currently no statistical agency in Japan that has announced its use. For further information on the aforementioned programs, please refer to the HP documentation on those developed by the U.S. Census Bureau. For further information, see the following link: https://www.census.gov/data/software/x13as.html.

we expect that our method has robustness properties when we handle small sample economic times series with non-stationary trend-cycle, and stationary seasonal and measurement error components because our method does not depend on the specific distribution as well as the dimension of the underlying random variables. See Kunitomo et al. (2020) for a comparison of small sample properties of the ML (maximum likelihood) and SIML methods for the non-stationary errors-in-variables models, and Nishimura et al. (2019) for an application of financial data smoothing. In financial time series, the number of data can be often huge in high-frequency market data and the main purpose of statistical analysis is usually different from the time analysis of macroeconomic data. The most important feature of the present procedure is that it may be applicable to small sample time series data with non-stationary trend-cycle, seasonal, and noise components and it has a statistical foundation based on the (real-valued) spectral decomposition of stochastic processes by a (real-valued) Fourier transformation, as we shall explain in this book.

The following is a brief overview of this book. In Chap. 2, we present some macroeconomic data that motivated the present study and define the non-stationary errors-in-variables models. In Chap. 3, the fundamental aspects of the SIML method are defined and the statistical basis of the method in the frequency domain is discussed. We discuss applications including an interpretation on the Müller-Watson method in econometrics and give two empirical applications of macro consumption in Japan. Chapter 4 is devoted to a discussion of the statistical properties of alternative estimation methods for non-stationary economic time series models as well as a characterization of the SIML estimation method. Subsequently, in Chap. 5, we introduce the concept of frequency regression and smoothing for noisy non-stationary economic time series, which represents an extension of the basic models presented in Chap. 3. In Chap. 6, we present an application of the SIML approach for construction of a macro consumption index based on monthly consumption data and quarterly data. Finally, in Chap. 7, we make some final comments and suggest possible extensions of the SIML filtering method. The mathematical derivations of theoretical results are presented in the appendices of each chapter. In order to facilitate comprehension of the methodologies presented in each chapter, we have occasionally included identical expressions for the reader's convenience.

References

Anderson TW (1984) Estimating linear statistical relationships. Annals of Statistics, 12:1–45

Anderson TW (2003) An Introduction to Multivariate Statistical Analysis, 3rd edn. Wiley

Baxter H, King R (1999) Measuring business cycles: approximate band-pass filters for economic time series. Rev Econ Stat 81–4:575–593

Brillinger D, Hatanaka M (1969) An harmonic analysis of nonstationary multivariate economic processes. Econometrica 35:131–141

Christiano L, Fitzgerald T (2003) The band pass filter. Int Econ Rev 44–2:435–465

Engle R, Granger CWJ (1987) Co-integration and error correction. Econometrica 55:251–276

Fuller WA (1987) Measurement errors models. John-Wiley

Granger CWJ, Hatanaka M (1964) Spectral analysis of economic time series. Princeton U.P

Harvey A, Trimbur T (2008) Trend estimation and the Hodorick-Prescott filter. J Japan Stat Soc 38–1:41–49

Johansen S (1995) Likelihood based inference in cointegrated vector autoregressive models, Oxford UP

Kitagawa G (2021) Introduction to time series modeling with applications in R, 2nd edn. CRC Press

Kunitomo N, Sato S (2017) Trend, seasonality and economic time series: the non-stationary errors-in-variables models. SDS-4, MIMS, Meiji University. http://www.mims.meiji.ac.jp/publications/2017-ds

Kunitomo N, Sato S, Kurisu D (2018) Separating information maximum likelihood estimation for high frequency financial data. Springer

Kunitomo N, Awaya N, Kurisu D (2020) Comparing Estimation methods of Non-stationary Errors-in-Variables Models. Jpn J Stat Data Sci 3:73–101. https://doi.org/10.1007/s42081-018-0017-3

Müller U, Watson M (2018) Long-run covariability. Econometrica 86–3:775–804

Nishimura GK, Sato S, Takahashi A (2019) Term structure models during the global financial crisis: a parsimonious text mining approach. Springer, Asia-Pacific Financial Markets

Chapter 2
Macro Examples and Non-stationary Errors-in-Variables Model

Abstract This chapter presents two empirical examples of macroeconomic data in Japan; real GDP and macro consumption series. These examples illustrate the main issues of our statistical analysis and the motivations behind our approach, which will be further elaborated upon in subsequent chapters. We then present the basic and general non-stationary errors-in-variables models, which are utilized extensively in the subsequent chapters.

2.1 Two Illustrative Examples

This subsection presents two illustrative examples of macroeconomic time series in Japan, which motivate the development of a new filtering method for non-stationary economic time series. As an initial illustrative example, we plot quarterly (real) GDP (1994Q1-2018Q2) and quarterly (real) consumption in Japan as the graph of typical two macroeconomic time series, as illustrated in Fig. 2.1.[1] This appears to be a simple example of linear regression, as frequently presented in introductory undergraduate-level courses. However, when we plot the time series sequences of these (original and seasonally unadjusted official) macroeconomic data published by ESRI, Cabinet Office of Japan as Fig. 2.2, it becomes evident that the situation is more complex than in Fig. 2.1. The two time series sets exhibit a number of distinct components, including clear trend-cycle components, seasonal fluctuations, and noise components. While many economists usually use the seasonally adjusted (published) data, which are generated through the X-12-ARIMA program across several government ministries, the precise impact of the filtering employed in the program is frequently unclear. The X-12-ARIMA program employs the univariate Reg-ARIMA model, which is a mixture of univariate seasonal ARIMA and linear regressions, and it decomposes univariate time series into the trend-cycle, seasonal, and noise

---

[1] It should be noted that in Japan, both the original quarterly series and the seasonally adjusted series of GDP and its major components are regularly published by ESRI (Economic and Social Research Institute), Cabinet office, Japan. It may differ from macroeconomic data in the U.S. in some aspect.

N. Kunitomo and S. Sato, *The SIML Filtering Method for Noisy Non-stationary Economic Time Series*, JSS Research Series in Statistics,
https://doi.org/10.1007/978-981-96-0882-9_2

Fig. 2.1 GDP versus consumption. (Data are the Quarterly real GDP (RGDP) and real consumption (RCONSUMPTION) between 1994Q1-2014Q3, which were published in 2015 by the Economic Social Research Institute (ESRI), Cabinet Office, Japan.)

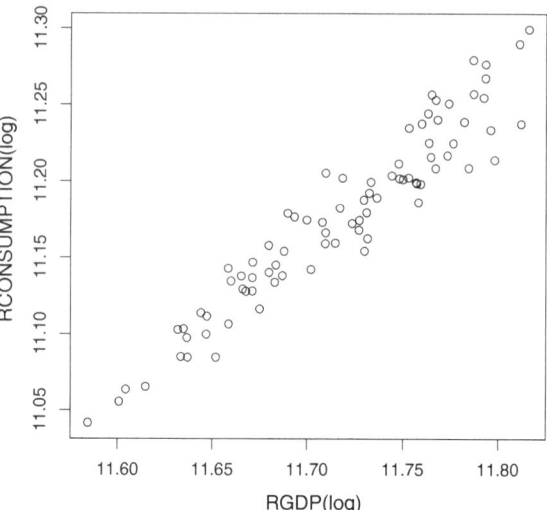

components as the standard procedure.[2] In contrast, the DECOMP program, which is explained by Kitagawa (2021), is another possible choice in Japan particularly, that employs the univariate AR model and the Kalman-filtering technique with AIC, which is based on the Gaussian likelihood. When each time series is handled using different filtering procedures (i.e., different ARIMA models or Reg-ARIMA models for instance), it may result in a fundamental problem in their interpretation when the focus is on the relationships among different non-stationary time series.

As the second example, Fig. 2.3 provides three different macro consumption data series (2002 January–2016 December), which are observed as monthly time series and widely used by economists in Japan to assess the prevailing macro-business conditions. The initial series is referred to *Kakei-Chosa* (the data on the monthly consumer-survey collected by Statistics Bureau, Ministry of Internal Affairs and Communications), the second series is *Shougyo-Doutai-Chosa* (the data on the monthly retail collected by Ministry of Economy, Trade and Industry (METI)) and the third series is *Dai-Sanji-Sangyo-Toukei* (the index data on tertiary industry constructed by METI). It is important to note that the data construction methods of these series based on sample surveys are intricate and vary between different government ministries, and each data reflects different aspects of macro consumption. Despite exhibiting comparable movements, we observe discernible discrepancies with respect to trend-cycle, seasonal, and irregular noise components. Then, it may be desirable to unify the monthly consumption series because we want to judge the business condition each month by just observing these data to evaluate the state of the Japanese macroeconomy and forming macroeconomic policy. The majority of economists in both the public and private sectors typically utilize seasonally adjusted data, which have been constructed from the original quarterly or monthly

---

[2] Further details of the X-12-ARIMA program can be found in the paper by Findley et al. (1998).

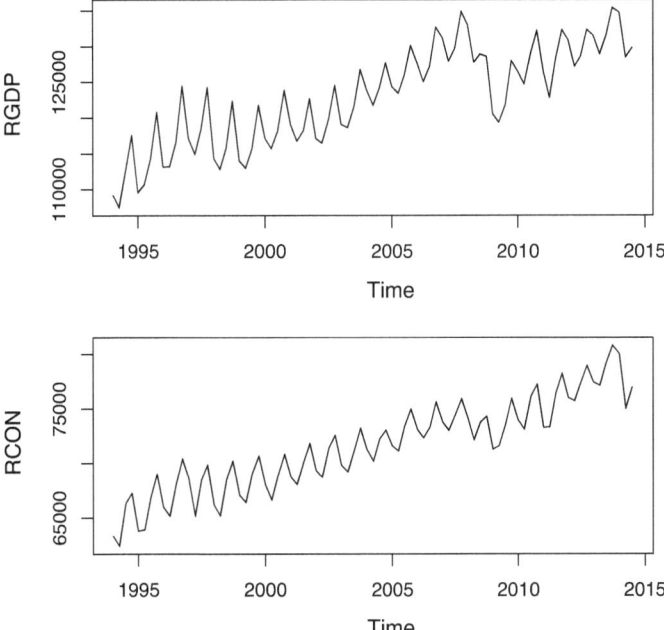

Fig. 2.2 GDP and consumption. (Data are the Quarterly real GDP (RGDP) and real consumption (RCON) between 1994Q1-2014Q3, which were published in 2015 by the Economic Social Research Institute (ESRI), Cabinet Office, Japan.)

(original) time series via the univariate X-12-ARIMA seasonal adjustment program. It is important to construct the monthly consumption index, which is consistent with the published quarterly macro consumption data as a part of GDP, which is usually reported with substantial time lags. It was one of motivations to develop our filtering method, and the resulting macro consumption index is discussed in Chap. 6.

Some econometricians make use of the multivariate (parametric) time series models such as the VAR (vector autoregressive model) for the analysis of macroeconomic data and the investigation of the relationships among macroeconomic variables. They may use the seasonally adjusted (official) data, but we need caution to use such data because most published official data are already filtered by the X-12-ARIMA program. When the dimension is more than 2, some difficulties handling trend-cycle, seasonal, and measurement errors simultaneously arise as a challenge. The necessity to process macroeconomic data in simple non-parametric manner prompted the development of multivariate non-stationary errors-in-variables models and a filtering method for the hidden state variables with measurement errors.

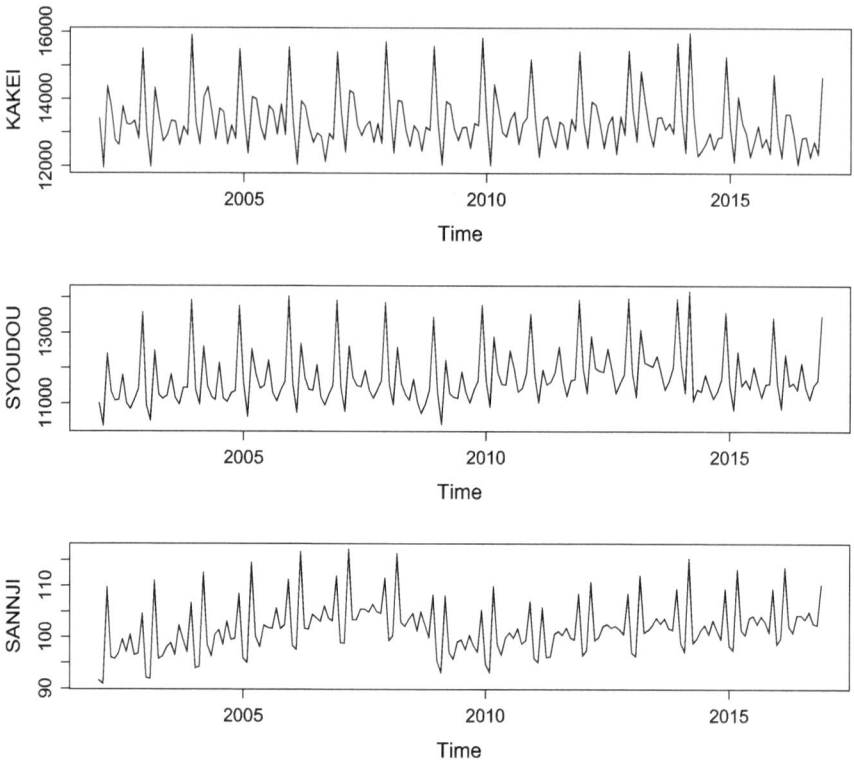

Fig. 2.3 Monthly consumption ceries. (Data are the monthly consumptions between 2002M1–2016M12, which were published in 2017 by the Statistics Bureau of Japan and Ministry of Economy, Trade and Industry (METI), Japan.)

2.2 Non-stationary Errors-in-Variables Model and SIML

2.2.1 A Simple Non-stationary Errors-in-Variables Model and the SIML Method

In this subsection we first introduce a simple non-stationary errors-in-variables model with trend and noise components and explain the SIML method. Then in the next subsection we shall investigate the general framework for non-stationary multivariate time series with trend-cycle, seasonal, and measurement error components.

Let y_{ji} be the i-th observation of the j-th time series at i for $i = 1, \ldots, n$; $j = 1, \ldots, p$. We set $\mathbf{y}_i = (y_{1i}, \ldots, y_{pi})^{'}$ be a $p \times 1$ vector and $\mathbf{Y}_n = (\mathbf{y}_i^{'})$ be an $n \times p$ matrix of observations and denote $\mathbf{y}_0$ as the initial $p \times 1$ vector and it is fixed. We consider the simple model that the underlying non-stationary trend is $\mathbf{x}_i' (= (x_{1i}, \ldots, x_{pi}), i = 1, \ldots, n)$, which is a sequence of non-stationary I(1) process that satisfies

$$\mathbf{x}_i = \mathbf{x}_{i-1} + \mathbf{v}_i^{(x)} \tag{2.1}$$

and the noise component is $\mathbf{v}_i' = (v_{1i}, \ldots, v_{pi})$, which is independent of $\mathbf{v}_i^{(x)}$. Then, we write

$$\mathbf{y}_i = \mathbf{x}_i + \mathbf{v}_i \quad (i = 1, \ldots, n) . \tag{2.2}$$

We use the notation of mathematical expectations as $\mathbf{E}(\mathbf{v}_i^{(x)}) = \mathbf{0}$, $\mathbf{E}(\mathbf{v}_i^{(x)} \mathbf{v}_i^{(x)'}) = \boldsymbol{\Sigma}_x$ (positive-semi-definite) for the noise components of trends, and $\mathbf{E}(\mathbf{v}_i) = \mathbf{0}$, $\mathbf{E}(\mathbf{v}_i \mathbf{v}_i) = \boldsymbol{\Sigma}_v$ (positive-semi-definite) for the measurement error components, respectively, in the simple case.

When each pair of vectors $\Delta \mathbf{x}_i$ $(= \mathbf{x}_i - \mathbf{x}_{i-1})$ and $\mathbf{v}_i$ are independently, identically, and normally distributed (i.i.d.) as $N_p(\mathbf{0}, \boldsymbol{\Sigma}_x)$ and $N_p(\mathbf{0}, \boldsymbol{\Sigma}_v)$, respectively, we have the observations of an $n \times p$ matrix $\mathbf{Y}_n = (\mathbf{y}_i')$ and set the $np \times 1$ random vector $(\mathbf{y}_1', \cdots, \mathbf{y}_n')'$. Given the initial condition $\mathbf{y}_0$ $(= \mathbf{x}_0)$, we have

$$\text{vec}(\mathbf{Y}_n) \sim N_{n \times p} \left(\mathbf{1}_n \cdot \mathbf{y}_0', \mathbf{I}_n \otimes \boldsymbol{\Sigma}_v + \mathbf{C}_n \mathbf{C}_n' \otimes \boldsymbol{\Sigma}_x \right) , \tag{2.3}$$

where $\mathbf{1}_n' = (1, \ldots, 1)$, $\mathbf{I}_n$ is the $n \times n$ identity matrix, and

$$\mathbf{C}_n = \begin{pmatrix} 1 & 0 & \cdots & 0 & 0 \\ 1 & 1 & 0 & \cdots & 0 \\ 1 & 1 & 1 & \cdots & 0 \\ 1 & \cdots & 1 & 1 & 0 \\ 1 & \cdots & 1 & 1 & 1 \end{pmatrix}_{n \times n} .$$

We use the $\mathbf{K}_n^*$-transformation that from $\mathbf{Y}_n$ to $\mathbf{Z}_n$ $(= (\mathbf{z}_k'))$ by

$$\mathbf{Z}_n = \mathbf{K}_n^* \left(\mathbf{Y}_n - \bar{\mathbf{Y}}_0 \right) , \mathbf{K}_n^* = \mathbf{P}_n \mathbf{C}_n^{-1}, \tag{2.4}$$

where $\bar{\mathbf{Y}}_0 = \mathbf{1}_n \cdot \mathbf{y}_0'$,

$$\mathbf{C}_n^{-1} = \begin{pmatrix} 1 & 0 & \cdots & 0 & 0 \\ -1 & 1 & 0 & \cdots & 0 \\ 0 & -1 & 1 & 0 & \cdots \\ 0 & 0 & -1 & 1 & 0 \\ 0 & 0 & 0 & -1 & 1 \end{pmatrix}_{n \times n} , \tag{2.5}$$

and

$$\mathbf{P}_n = (p_{jk}^{(n)}) , \ p_{jk}^{(n)} = \sqrt{\frac{2}{n + \frac{1}{2}}} \cos \left[\frac{2\pi}{2n + 1} (k - \frac{1}{2})(j - \frac{1}{2}) \right] . \tag{2.6}$$

Using the spectral decomposition $\mathbf{C}_n^{-1}\mathbf{C}_n^{'-1} = \mathbf{P}_n\mathbf{D}_n\mathbf{P}_n$ and $\mathbf{D}_n$ is a diagonal matrix with the k-th element $d_k = 2[1 - \cos(\pi(\frac{2k-1}{2n+1}))]$ $(k = 1, \ldots, n)$ (see Lemma A.1 in the Appendix of this chapter), we write

$$a_{kn}^* (= d_k) = 4\sin^2\left[\frac{\pi}{2}\left(\frac{2k-1}{2n+1}\right)\right] \quad (k = 1, \ldots, n). \tag{2.7}$$

Then, the separating information maximum likelihood (SIML) estimator of $\boldsymbol{\Sigma}_x$ in (2.2) can be defined by

$$\mathbf{G}_m = \hat{\boldsymbol{\Sigma}}_{x, SIML} = \frac{1}{m_n}\sum_{k=1}^{m_n}\mathbf{z}_k\mathbf{z}_k', \tag{2.8}$$

where we set $m = m_n = [n^{\alpha}]$ $(0 < \alpha < 1)$.

We need to use m terms in n and m depends on n. We take m is smaller than n substantially in (2.8), and the reason for this will become clear from the frequency domain analysis in Chap. 3.

For the estimation of the variance-covariance matrix $\boldsymbol{\Sigma}_v$, some estimators will be discussed in Chaps. 3 and 4. As we will see in later chapters, the SIML estimation method is quite robust even when $\mathbf{v}_i^{(x)}$ and $\mathbf{v}_i$ are non-Gaussian stationary processes, and they are serially correlated.

2.2.2 General Non-stationary Errors-in-Variables Model

In order to investigate non-stationary trend-cycle and stationary seasonal and measurement error components, we consider the general non-stationary multivariate errors-in-variables model[3]

$$\mathbf{y}_i = \mathbf{x}_i + \mathbf{s}_i + \mathbf{v}_i \quad (i = 1, \ldots, n). \tag{2.9}$$

We assume that each components x_i, v_i, and s_i $(i = 1, \ldots, n)$ are mutually independent in the following analysis. We explain the general model in three steps.

(i) The trend-cycle factor $\mathbf{x}_i$ $(i = 1, \ldots, n)$ is a sequence of non-stationary I(1) process that satisfies

$$\Delta\mathbf{x}_i = (1 - \mathcal{L})\mathbf{x}_i = \mathbf{v}_i^{(x)}, \tag{2.10}$$

with the lag operator $\mathcal{L}\mathbf{x}_i = \mathbf{x}_{i-1}$, $\Delta = 1 - \mathcal{L}$,

---

[3] It is possible to use the log-transformed data for multiplicative models. It is known that the standard model in X-12-ARIMA is multiplicative, even though it uses moving averages, for instance. We also extend (2.9) in Chap. 5 to include some explanatory variables.

$$\mathbf{v}_i^{(x)} = \sum_{j=0}^{\infty} \mathbf{C}_j^{(x)} \mathbf{e}_{i-j}^{(x)} , \tag{2.11}$$

and $\mathbf{e}_i^{(x)}$ is a sequence of i.i.d. random vectors with $\mathbf{E}(\mathbf{e}_i^{(x)}) = \mathbf{0}$ and $\mathbf{E}(\mathbf{e}_i^{(x)} \mathbf{e}_i^{(x)'}) = \boldsymbol{\Sigma}_e^{(x)}$ (positive-semi-definite). The $p \times p$ coefficient matrices $\mathbf{C}_j^{(x)} (= c_{kl}^{(x)}(j))$ are absolutely summable and $\|\mathbf{C}_j^{(x)}\| = O(\rho^j)$, where $0 \le \rho < 1$ and $\|\mathbf{C}_j^{(x)}\| = \max_{k,l=1,\dots,p} |c_{kl}^{(x)}(j)|$.

(ii) The random noise (or measurement errors) vectors $\mathbf{v}_i$ ($i = 1, \dots, n$) are a sequence of stationary I(0) process with

$$\mathbf{v}_i = \sum_{j=0}^{\infty} \mathbf{C}_j^{(v)} \mathbf{e}_{i-j}^{(v)} , \tag{2.12}$$

where the $p \times p$ coefficient matrices $\mathbf{C}_j^{(v)}$ are absolutely summable and $\|\mathbf{C}_j^{(v)}\| = O(\rho^j)$, where $0 \le \rho < 1$ and $\mathbf{e}_i^{(v)}$ is a sequence of i.i.d. random vectors with $\mathbf{E}(\mathbf{e}_i^{(v)}) = \mathbf{0}$, $\mathbf{E}(\mathbf{e}_i^{(v)} \mathbf{e}_i^{(v)'}) = \boldsymbol{\Sigma}_e^{(v)}$ (positive definite).

(iii) We take a positive integer s ($s > 1$), N, and $n = sN$ for the resulting simplicity of exposition and arguments. The seasonal factor $\mathbf{s}_i$ ($i = 1, \dots, n$) is a sequence of stationary process,[4] which satisfies

$$\mathbf{s}_i = \sum_{j=0}^{\infty} \mathbf{C}_{sj}^{(s)} \mathbf{e}_{i-sj}^{(s)} , \tag{2.13}$$

where the lag operator is defined by $\mathscr{L}^s \mathbf{s}_i = \mathbf{s}_{i-s}$ ($s \ge 2$), and $\mathbf{e}_i^{(s)}$ is a sequence of i.i.d. random vectors with $\mathbf{E}(\mathbf{e}_i^{(s)}) = \mathbf{0}$ and $\mathbf{E}(\mathbf{e}_i^{(s)} \mathbf{e}_i^{(s)'}) = \boldsymbol{\Sigma}_e^{(s)}$ (a non-negative definite matrix). The $p \times p$ coefficient matrices $\mathbf{C}_j^{(s)}$ are absolutely summable and $\|\mathbf{C}_j^{(s)}\| = O(\rho^j)$, where $0 \le \rho < 1$.

(iv) From our conditions imposed in this subsection, the variance-covariance matrices $\mathbf{E}(\mathbf{v}_i^{(x)} \mathbf{v}_i^{(x)'}) = \boldsymbol{\Sigma}_x$, $\mathbf{E}(\mathbf{v}_i \mathbf{v}_i') = \boldsymbol{\Sigma}_v$, and $\mathbf{E}(\mathbf{s}_i \mathbf{s}_i') = \boldsymbol{\Sigma}_s$ are well-defined. Furthermore, we can define the autocovariance-matrices by $\mathbf{E}(\mathbf{v}_{h+i}^{(x)} \mathbf{v}_i^{(x)'}) = \boldsymbol{\Gamma}_{\Delta x}(h)$, $\mathbf{E}(\mathbf{v}_{h+i} \mathbf{v}_i') = \boldsymbol{\Gamma}_v(h)$, and $\mathbf{E}(\mathbf{s}_{h+i} \mathbf{s}_i') = \boldsymbol{\Gamma}_s(h)$ for any integer h.

Let $\mathbf{f}_{\Delta x}(\lambda)$, $\mathbf{f}_v(\lambda)$, and $\mathbf{f}_s(\lambda)$ be the spectral density (see Appendix of this chapter for the definition) of $p \times p$ matrices of $\Delta \mathbf{x}_i$, $\mathbf{v}_i$, and $\mathbf{s}_i$ ($i = 1, \cdots, n$), respectively, which are complex-valued functions. They are given by

$$\mathbf{f}_{\Delta x}(\lambda) = \left(\sum_{j=0}^{\infty} \mathbf{C}_j^{(x)} e^{2\pi i \lambda j} \right) \boldsymbol{\Sigma}_e^{(x)} \left(\sum_{j=0}^{\infty} \mathbf{C}_j^{(x)'} e^{-2\pi i \lambda j} \right), \left(-\frac{1}{2} \le \lambda \le \frac{1}{2} \right), \tag{2.14}$$

---

[4] In this book, we consider the case when $\mathbf{s}_i$ is stationary. When we assume that $\Delta \mathbf{s}_i$ is stationary (Kitagawa 2021, Chap. 12 for instance), some arguments in the following sections would go through.

$$\mathbf{f}_v(\lambda) = \left(\sum_{j=0}^{\infty} \mathbf{C}_j^{(v)} e^{2\pi i \lambda j} \right) \mathbf{\Sigma}_e^{(v)} \left(\sum_{j=0}^{\infty} \mathbf{C}_j^{(v)\prime} e^{-2\pi i \lambda j} \right), \left(-\frac{1}{2} \le \lambda \le \frac{1}{2} \right), \quad (2.15)$$

and

$$\mathbf{f}_s(\lambda) = \left(\sum_{j=0}^{\infty} \mathbf{C}_{sj}^{(s)} e^{2\pi i \lambda s j} \right) \mathbf{\Sigma}_e^{(s)} \left(\sum_{j=0}^{\infty} \mathbf{C}_{sj}^{(s)\prime} e^{-2\pi i \lambda s j} \right) \left(-\frac{1}{2} \le \lambda \le \frac{1}{2} \right), \quad (2.16)$$

where we set $\mathbf{C}_0^{(x)} = \mathbf{C}_0^{(v)} = \mathbf{C}_0^{(s)} = \mathbf{I}_p$ for normalization and $i^2 = -1$ (see Chap. 7 of Anderson (1971)).

Then the $p \times p$ spectral density matrix of the transformed vector process of difference series $\Delta \mathbf{y}_i \ (= \mathbf{y}_i - \mathbf{y}_{i-1})$ can be represented as

$$\mathbf{f}_{\Delta y}(\lambda) = \mathbf{f}_{\Delta x}(\lambda) + (1 - e^{2\pi i \lambda})[\mathbf{f}_s(\lambda) + \mathbf{f}_v(\lambda)](1 - e^{-2\pi i \lambda}) \ . \quad (2.17)$$

We denote the long-run variance-covariance matrices of trend-cycle and noise components for $g, h = 1, \ldots, p$ as

$$\mathbf{\Sigma}_x = \mathbf{f}_{\Delta x}(0) \ (= (\sigma_{gh}^{(x)})), \ \mathbf{\Sigma}_v = f_v(0) \ = (\sigma_{gh}^{(v)}), \quad (2.18)$$

respectively.

One important often overlooked issue in practical applications is that when applying the differencing procedure to non-stationary time series and using the standard statistical method for multivariate stationary time series, there is no guarantee of preserving the relationships among the original time series through the transformations. Although Engle and Granger (1987) and Johansen (1995) noticed this problem, they did not consider the frequency domain aspect with seasonality and measurement errors. The SIML filtering approach presented in this book may shed a new light on the relationships among the time domain and frequency decompositions of non-stationary multivariate time series.

Appendix of this Chapter: Mathematical Supplements

This appendix presents the key relations used in Sect. 2.2. In (2.4)–(2.7), we have used the characteristic roots and vectors, which are the most important relations to our filtering approach for non-stationary economic time series. We also give the definition of the spectral density matrix, which is the key to our interpretation in the frequency domain. It is in the standard time series textbooks such as Andrson (1971) and Brockwell and Davis (1990) on the frequency domain analysis in more detail.

Lemma 2.1 *(i) Define an $n \times n$ matrix $\mathbf{A}_n$ by*

$$\mathbf{A}_n = \frac{1}{2} \begin{pmatrix} 1 & 1 & 0 & \cdots & 0 \\ 1 & 0 & 1 & \cdots & 0 \\ 0 & 1 & 0 & 1 & \cdots \\ 0 & 0 & \cdots & 0 & 1 \\ 0 & \cdots & 0 & 1 & 0 \end{pmatrix} . \tag{2.19}$$

Then, $\cos \pi \left(\frac{2k-1}{2n+1} \right)$ $(k = 1, \ldots, n)$ are eigen-values of $\mathbf{A}_n$ and the eigen-vectors are

$$\begin{bmatrix} \cos[\pi (\frac{2k-1}{2n+1}) \frac{1}{2}] \\ \cos[\pi (\frac{2k-1}{2n+1}) \frac{3}{2}] \\ \vdots \\ \cos[\pi (\frac{2k-1}{2n+1})(n - \frac{1}{2})] \end{bmatrix} \quad (k = 1, \ldots, n). \tag{2.20}$$

(ii) We have the spectral decomposition

$$\mathbf{C}_n^{-1} \mathbf{C}_n'^{-1} = \mathbf{P}_n \mathbf{D}_n \mathbf{P}_n = 2\mathbf{I}_n - 2\mathbf{A}_n , \tag{2.21}$$

where $\mathbf{D}_n$ is a diagonal matrix with the k-th element

$$d_k = 2 \left[1 - \cos(\pi (\frac{2k-1}{2n+1})) \right] \quad (k = 1, \ldots, n) , \tag{2.22}$$

and $\mathbf{C}_n^{-1}$ and $\mathbf{P}_n$ are defined by (2.5) and (2.6).

Lemma 2.2 *(i) Let $\mathbf{A}_n = (a_{ij})$ in (2.19) $(i, j = 1, \ldots, n)$ and an $n \times 1$ vector $\mathbf{x} = (x_t)$ $(t = 1, \ldots, n)$ satisfying $\mathbf{A}_n \mathbf{x} = \lambda \mathbf{x}$. Then*

$$\frac{x_1 + x_2}{2} = \lambda x_1, \tag{2.23}$$

$$\frac{x_{t-1} + x_{t+1}}{2} = \lambda x_t (t = 2, \ldots, n - 1), \tag{2.24}$$

$$\frac{1}{2} x_{n-1} = \lambda x_n. \tag{2.25}$$

Let ξ_i $(i = 1, 2)$ be the solutions of $\xi^2 - 2\lambda\xi + 1 = 0$. Because $2\lambda = \xi_1 + \xi_2$ and $\xi_1 \xi_2 = 1$, we have the solution of (2.24) as $x_t = c_1 \xi_1^t + c_2 \xi_1^{-t}$ $(t = 1, \cdots, n)$ and c_i $(i = 1)$ are real constants. Then (2.23) implies

$$0 = c_1 \xi_1 + c_2 \xi_1^{-1} + c_1 \xi_1^2 + c_2 \xi_1^{-2} - (\xi_1 + \xi_1^{-1})(c_1 \xi_1 + c_2 \xi_1^{-1})$$
$$= (\xi_1 - 1)(c_1 - c_2 \xi_1^{-1}).$$

Because $\xi = 1$ *cannot be a solution (i.e., it leads to a contradiction), we find that* $c_2 = c_1\xi_1$. *Then we find that* $x_t = c_1[\xi_1^t + \xi_1^{-(t-1)}]$ *and (2.25) implies* $\xi_1^{2n+1} = -1$. *Then we find that* $\xi_1 = \exp[i\pi(2k-1)/(2n+1)]$, $\xi_2 = \exp[-\pi(2k-1)/(2n+1)]$, *and*

$$\lambda_k = \cos[\pi\frac{2k-1}{2n+1}] \ (k = 1, \ldots, n). \tag{2.26}$$

By taking $c_1 = (1/2)\xi_1^{-1/2}$, *the elements of the characteristic vectors of* $\mathbf{A}_n$ *with* $\cos[\pi(2k-1)/(2n+1)]$ *are*

$$x_t = \frac{1}{2}\left[\xi_1^{t-1/2} + \xi_1^{-(t-1/2)}\right] = \cos\left[\pi\frac{2k-1}{2n+1}(t - \frac{1}{2})\right]. \tag{2.27}$$

(ii) The rest of the proof involves the standard arguments of spectral decomposition in linear algebra. Q.E.D.

Definition of A.1[5]: For a $p \times 1$ stationary process $\mathbf{r}_i \ (= (r_{ji}), \ j = 1, \ldots, p)$, define the covariance function $\boldsymbol{\Gamma}(h) = \mathbf{E}[(\mathbf{r}_{i+h} - \boldsymbol{\mu})(\mathbf{r}_i' - \boldsymbol{\mu})]$ with $i = 0, \pm 1, \pm 2, \ldots$, where $\boldsymbol{\mu} = \mathbf{E}[\mathbf{r}_i]$ and $\sum_{h=-\infty}^{\infty} |\gamma_{jk}(h)| < \infty \ (j, k = 1, \ldots, p)$. Then, the $p \times p$ spectral density is defined by

$$f(\lambda) = \sum_{h=-\infty}^{\infty} \exp(-2\pi\lambda i)\boldsymbol{\Gamma}(h) \ \left(-\frac{1}{2} \leq \lambda \leq \frac{1}{2}\right), \tag{2.28}$$

where $i^2 = -1$.

References

Anderson TW (1971) The statistical analysis of time series. Wiley

Brockwell P, Davis R (1990) Time series: theory and methods, 2nd edn. Wiley

Engle R, Granger CWJ (1987) Co-integration and error correction. Econometrica 55:251–276

Findley D, Monsell B, Bell W, Otto M, Chen B (1998) New capabilities and methods of the X-12-ARIMA seasonal adjustment program. J Bus Econ Stat 16:127–177

Johansen S (1995) Likelihood based inference in cointegrated vector autoregressive models, Oxford UP

Kitagawa G (2021) Introduction to time series modeling with applications in R, 2nd edn. CRC Press

---

[5] Our notation of spectral density is slightly different from the standard notation used in Anderson (1971). Let $\mu = 2\pi\lambda$ and $f^A(\mu) \ (-\pi \leq \mu \leq \pi)$ be the spectral density in Chap. 7 of Anderson (1971). Then, $f(\lambda) = 2\pi f^A(\mu) \ (-\frac{1}{2} \leq \lambda \leq \frac{1}{2})$. The present definition of spectral density corresponds to $2\pi f^A(\mu)$ in some literature.

Chapter 3
The SIML Filtering Method

Abstract We introduce the SIML filtering method of hidden random variables of trend-cycle, seasonal, and measurement errors components and propose a method to handle macroeconomic time series. We develop the asymptotic theory based on the frequency domain analysis for non-stationary time series. We illustrate some applications and analyses of macro consumption data in Japan. We also discuss the relation of our method to the one by Muller and Watson (Econometrica 86–3:775–804, 2018) in econometrics.

3.1 The SIML Filtering Method

3.1.1 Basic Filtering

We consider $p \times 1$ vectors $\mathbf{y}_i$ $(i = 1, \ldots, n)$ in (2.9)–(2.13) and set the $n \times p$ random matrix $\mathbf{Y} = (\mathbf{y}_i')$. We investigate the general filtering procedure based on the $\mathbf{K}_n^*$-transformation in (2.4). When we interpret that the elements of the resulting $n \times p$ random matrix $\mathbf{Z}_n$ take real values in the frequency domain, it is easy to understand their roles as we will explain shortly in this chapter. Since $\mathbf{P}_n$ is a kind of real-valued discrete Fourier transform, vectors $\mathbf{z}_k$ $(k = 1, \ldots, n)$ in $\mathbf{Z}_n$ are asymptotically uncorrelated (see Sect. 3.2.2). We investigate the partial inversion of the transformed orthogonal processes. Let an $n \times p$ matrix

$$\hat{\mathbf{X}}_n(Q) = \mathbf{C}_n \mathbf{P}_n \mathbf{Q}_n \mathbf{P}_n \mathbf{C}_n^{-1} (\mathbf{Y}_n - \bar{\mathbf{Y}}_0) \tag{3.1}$$

and

$$\mathbf{Z}_n = \mathbf{P}_n \mathbf{C}_n^{-1} (\mathbf{Y}_n - \bar{\mathbf{Y}}_0), \; \mathbf{Y}_n = \bar{\mathbf{Y}}_0 + \mathbf{X}_n^* + \mathbf{S}_n + \mathbf{V}_n, \tag{3.2}$$

where $\mathbf{X}_n^* = (\mathbf{x}_i^{*'})$, $\mathbf{S}_n = (\mathbf{s}_i')$, and $\mathbf{V}_n = (\mathbf{v}_i')$ are $n \times p$ matrices, $\mathbf{x}_i^* = \mathbf{x}_i - \mathbf{x}_0$ $(i = 1, \ldots, n)$ and $\mathbf{x}_0 = \mathbf{y}_0$ is the initial vector. The $n \times n$ matrix $\mathbf{P}_n$ is given by (2.6) and $\mathbf{Q}_n$ is an $n \times n$ filtering matrix, which is defined shortly.

The stochastic process $\mathbf{Z}_n$ is the orthogonal decomposition of the original time series $\mathbf{Y}_n$ in the frequency domain. Because $\mathbf{Y}_n$ consist of non-stationary time series,

N. Kunitomo and S. Sato, *The SIML Filtering Method for Noisy Non-stationary Economic Time Series*, JSS Research Series in Statistics,
https://doi.org/10.1007/978-981-96-0882-9_3

we need a special form of transformation $\mathbf{K}_n^*$ in (2.4). We give explicit forms of two examples, including the trend-cycle filtering and the band filtering procedures. Although there can be many possible filtering procedures within our general framework, it is useful to discuss linear filtering procedures.

Let an $n \times n$ diagonal matrix

$$\mathbf{Q}_n = \sum_{i=1}^{n} w_i^{(n)} \mathbf{e}_i^{(n)} \mathbf{e}_i^{(n)'} \tag{3.3}$$

and $\mathbf{e}_i^{(n)} = (0, \ldots, 1, \ldots, 0)'$ $(i = 1, \ldots, n)$ are the unit vectors $(\mathbf{e}_i^{(n)'} \mathbf{e}_i^{(n)} = 1)$ and $w_i^{(n)}$ $(i = 1, \ldots, n)$ are some non-negative constants.

We start with the case when $w_i^{(n)} = 1$ $(i = 1, \ldots, n)$ and we have the identity matrix $\mathbf{Q}_n = \mathbf{I}_n$. Then we find $\mathbf{C}_n \mathbf{P}_n \mathbf{Q}_n \mathbf{P}_n \mathbf{C}_n^{-1} = \mathbf{I}_n$. There can be useful cases and we present two cases as the trend-cycle filtering and the band filtering by choosing $w_{i,n} = 1$ or 0 for some $i's$. Although the first example could be regarded as a special case of the second one, the analysis of trend-cycle component in the first example has an important role for analyzing economic time series.

(i) **Example 3.1 : Trend-Cycle Filtering**: Let an $m \times n$ $(m < n)$ choice matrix $\mathbf{J}_m = (\mathbf{I}_m, \mathbf{O})$, and let also $n \times p$ matrix

$$\hat{\mathbf{X}}_n^{(m)} = \mathbf{C}_n \mathbf{P}_n \mathbf{J}_m' \mathbf{J}_m \mathbf{P}_n \mathbf{C}_n^{-1} (\mathbf{Y}_n - \bar{\mathbf{Y}}_0) \tag{3.4}$$

and an $n \times n$ matrix $\mathbf{Q}_n = \mathbf{J}_m' \mathbf{J}_m$.

We construct an estimator of $n \times p$ hidden state matrix $\mathbf{X}_n^*$ only in the lower frequency parts by using the inverse transformation of $\mathbf{Z}_n$ and deleting the estimated seasonal and noise parts. We denote the hidden trend-cycle state based on m frequencies as

$$\mathbf{X}_n^{(m)} = \mathbf{C}_n \mathbf{P}_n \mathbf{J}_m' \mathbf{J}_m \mathbf{P}_n \mathbf{C}_n^{-1} \mathbf{X}_n^*. \tag{3.5}$$

This quantity is different from $\mathbf{X}_n^*$ because $\mathbf{x}_i$ $(i = 1, \ldots, n)$ in (2.9) and (2.10) contains not only the trend-cycle component of $\mathbf{y}_i$ $(i = 1, \ldots, n)$, but also the noise component in the frequency domain, which is different from the measurement noise component $\mathbf{v}_i$ $(i = 1, \ldots, n)$ in (2.9)–(2.13). We try to estimate the trend-cycle component of $\mathbf{x}_i$ by using (3.4) and recover the trend-cycle component of $\mathbf{X}_n$ near at the zero frequency because the effects of differenced measurement error noises $(\mathbf{v}_i - \mathbf{v}_{i-1})$ are negligible around at zero frequency. This method differs from some existing procedures that consider the decomposition of time series only in the time domain. Our arguments can be justified by using the frequency decomposition of $\mathbf{y}_i$ and $\mathbf{r}_i^{(n)} = \Delta \mathbf{y}_i$ $(= \mathbf{y}_i - \mathbf{y}_{i-1})$ when the initial vector $\mathbf{y}_0$ is being fixed. (We shall discuss this issue in Sect. 3.2.)

We partition $\mathbf{P}_n$ into $[m + (n - m)] \times [m + (n - m)]$ matrices as

$$\mathbf{P}_n = \begin{pmatrix} \mathbf{P}_{11} & \mathbf{P}_{12} \\ \mathbf{P}_{21} & \mathbf{P}_{22} \end{pmatrix}$$

and then

$$\mathbf{P}_n \mathbf{J}'_m \mathbf{J}_m \mathbf{P}_n = \begin{pmatrix} \mathbf{P}_{11} \\ \mathbf{P}_{21} \end{pmatrix} (\mathbf{P}_{11}, \mathbf{P}_{12}) = \mathbf{I}_n - \begin{pmatrix} \mathbf{P}_{12} \\ \mathbf{P}_{22} \end{pmatrix} (\mathbf{P}_{21}, \mathbf{P}_{22}) .$$

After straightforward calculations (see the Appendix of this chapter for the derivation), the (j, j')-th element of $\mathbf{A}_n = \mathbf{P}_n \mathbf{J}'_m \mathbf{J}_m \mathbf{P}_n \ (= (a^{(n,m)}_{j,j'}))$ is given by

$$a^{(n,m)}_{j,j} = \frac{2m}{2n+1} + \frac{1}{2n+1} \left[\frac{\sin \frac{2m\pi}{2n+1}(2j-1)}{\sin \frac{\pi}{2n+1}(2j-1)} \right],$$

$$a^{(n,m)}_{j,j'} = \frac{1}{2n+1} \left[\frac{\sin \frac{2m\pi}{2n+1}(j+j'-1)}{\sin \frac{\pi}{2n+1}(j+j'-1)} + \frac{\sin \frac{2m\pi}{2n+1}(j-j')}{\sin \frac{\pi}{2n+1}(j-j')} \right] \quad (j \neq j'). (3.6)$$

It is possible to evaluate MSE of statistical state vector estimation. Let $\hat{\mathbf{X}}^{(m)}_n = (\hat{\mathbf{x}}^{(m)}_{ni})$ ($\hat{\mathbf{x}}^{(m)}_{ni}$ is an $n \times 1$ vector, $i = 1, \ldots, p$) and $\mathbf{X}^{(m)}_n = (\mathbf{x}^{(m)}_{ni})$ ($\mathbf{x}^{(m)}_{ni}$ is an $n \times 1$ vector for $i = 1, \ldots, p$). By decomposing $\mathbf{Y}_n - \bar{\mathbf{Y}}_0 = \mathbf{X}^*_n + \mathbf{S}_n + \mathbf{V}_n$ and $\hat{\mathbf{X}}^{(m)}_n - \mathbf{X}^*_n = \mathbf{C}_n \mathbf{P}_n \mathbf{J}'_m \mathbf{J}_m \mathbf{P}_n \mathbf{C}^{-1}_n (\mathbf{S}_n + \mathbf{V}_n)$, we find

$$\mathcal{E}[(\hat{\mathbf{X}}^{(m)}_{ni} - \mathbf{X}^{(m)}_{ni})(\hat{\mathbf{X}}^{(m)}_{nj} - \mathbf{X}^{(m)}_{nj})'] = \mathbf{K}^{*-1}_n \mathbf{Q}_n \mathbf{K}^*_n \mathbf{\Gamma}^{(s+v)}_n (i, j) \mathbf{K}^{*'}_n \mathbf{Q}_n \mathbf{K}^{*'-1}_n,$$

where $\mathbf{K}^*_n = \mathbf{P}_n \mathbf{C}^{-1}_n$, $\mathbf{\Gamma}^{(s+v)}_n (i, j)$ is the $n \times n$ variance-covariance matrix of $\mathbf{s}_{ni} + \mathbf{v}_{ni}$ and $\mathbf{s}_{nj} + \mathbf{v}_{nj}$ ($n \times 1$ vectors for $i, j = 1, \ldots, p$), and $\mathbf{s}_{ni}$ and $\mathbf{v}_{ni}$ ($n \times 1$ vectors) are the i-th and j-th column vectors of $\mathbf{S}_n$ and $\mathbf{V}_n$, respectively.

Since (3.5) depends on $\mathbf{x}_i$, (3.1) minimizes the MSE with respect to unknown state vector $\mathbf{x}_i$ to estimate (3.5) and it is optimal in this sense.

(ii) **Example 3.2: Band Filtering**: We consider a general filtering based on the $\mathbf{K}^*_n$ transformation in (3.4) and use the inversion of some frequency parts of the random matrix $\mathbf{Z}_n$. An important example is the analysis of seasonal frequencies in the discrete time series and we take $s \ (> 1)$ being a positive integer as the seasonal lag.

Let an $m_2 \times [m_1 + m_2 + (n - m_1 - m_2)]$ choice matrix $\mathbf{J}_{m_1, m_2} = (\mathbf{O}, \mathbf{I}_{m_2}, \mathbf{O})$ (we take $m_1 + m_2 < n$), and let also $n \times p$ matrix

$$\hat{\mathbf{X}}^{(m_1, m_2)}_n = \mathbf{C}_n \mathbf{P}_n \mathbf{J}'_{m_1, m_2} \mathbf{J}_{m_1, m_2} \mathbf{P}_n \mathbf{C}^{-1}_n (\mathbf{Y}_n - \bar{\mathbf{Y}}_0) \tag{3.7}$$

and an $n \times n$ matrix $\mathbf{Q}_n = \mathbf{J}_{m_1,m_2}' \mathbf{J}_{m_1,m_2}$.

When we have a particular seasonal frequency s (> 1), for instance, we can take $m_1 = [2n/s] - h$ and $m_2 = 2h + 1$; ($h \geq 1$). We set $s = 4$ for quarterly data and $s = 12$ for monthly data. (We need to have several seasonal frequencies in data analysis of discrete time series and it was pointed out by Granger and Hatanaka (1964) already.)

As in the trend-cycle filtering problem, (3.7) is the SIML filtering value for

$$\mathbf{X}_n^{(m_1,m_2)} = \mathbf{C}_n \mathbf{P}_n \mathbf{J}_{m_1,m_2}' \mathbf{J}_{m_1,m_2} \mathbf{P}_n \mathbf{C}_n^{-1} (\mathbf{X}_n^* + \mathbf{S}_n), \tag{3.8}$$

and it is an estimate of some frequency components of $\mathbf{x}_i + \mathbf{s}_i$ ($i = 1, \ldots, n$) in (2.9), (2.10), and (2.13). The filtering problem becomes difficult because some frequency component of $\mathbf{y}_i$ includes not only some component $\mathbf{x}_i$ at the same frequency, but also some component of the measurement errors at the same frequency.

After straightforward calculations (see the Appendix of this chapter) in this case, the (j, j')-th element of $\mathbf{A}_n = \mathbf{P}_n \mathbf{J}_{m_1,m_2}' \mathbf{J}_{m_1,m_2} \mathbf{P}_n$ ($= (a_{j,j'}^{(n,m_1,m_2)})$) is given by

$$a_{j,j}^{(n,m_1,m_2)} = \frac{2m_2}{2n+1} + \frac{1}{2n+1} \left[\frac{\sin \frac{2(m_1+m_2)\pi}{2n+1}(2j-1) - \sin \frac{2(m_1)\pi}{2n+1}(2j-1)}{\sin \frac{\pi}{2n+1}(2j-1)} \right], \tag{3.9}$$

$$a_{j,j'}^{(n,m_1,m_2)} = \frac{1}{2n+1} \left[\frac{\sin \frac{2(m_1+m_2)\pi}{2n+1}(j+j'-1) - \sin \frac{2(m_1)\pi}{2n+1}(j+j'-1)}{\sin \frac{\pi}{2n+1}(j+j'-1)} \right.$$
$$\left. + \frac{\sin \frac{2(m_1+m_2)\pi}{2n+1}(j-j') - \sin \frac{2(m_1)\pi}{2n+1}(j-j')}{\sin \frac{\pi}{2n+1}(j-j')} \right] \quad (j \neq j').$$

When $m_1 = 0$ and $m_2 = m$, the resulting formula reduces to the trend-cycle filtering case. There are existing filtering procedures having the frequency domain interpretation, but it seems that our procedure differs from some existing literature because of (3.1) and (3.2).

It is also possible to evaluate MSE of the state vector estimation in the same way as Example 3.1. Let $\hat{\mathbf{X}}_n^{(m_1,m_2)}$ be an estimate of the hidden state as
$\mathbf{X}_n^{(m_1,m_2)} = \mathbf{C}_n \mathbf{P}_n \mathbf{J}_{m_1,m_2,n}' \mathbf{J}_{m_1,m_2,n} \mathbf{P}_n \mathbf{C}_n^{-1} (\mathbf{X}_n^* + \mathbf{S}_n)$.

By using the similar calculation as Example 3.1 and the notation of $\mathbf{\Gamma}_n^{(v)}(i, j)$ ($i, j = 1, \ldots, p$), we find

$$\mathcal{E}[(\hat{\mathbf{X}}_{ni}^{(m_1,m_2)} - \mathbf{X}_{ni}^{(m_1,m_2)})(\hat{\mathbf{X}}_{nj}^{(m_1,m_2)} - \mathbf{X}_{nj}^{(m_1,m_2)})')] \tag{3.10}$$
$$= \mathbf{K}_n^{*-1} \mathbf{Q}_n \mathbf{K}_n^* \mathbf{\Gamma}_n^{(v)}(i, j) \mathbf{K}_n^{*'} \mathbf{Q}_n \mathbf{K}_n^{*'-1},$$

where $\mathbf{\Gamma}_n^{(v)}(i, j)$ is the $n \times n$ variance-covariance matrix of $\mathbf{v}_{ni}$ and $\mathbf{v}_{nj}$.

For a particular frequency such as seasonal component, we may use an alternative notation $m_1 = [\frac{2n}{s}] - h$, $m_2 = [\frac{2n}{s}] + h$ by using an $(m_2 - m_1) \times n$ choice matrix $\mathbf{J}_{m_1,m_2} = (\mathbf{O}, \mathbf{I}_{m_2-m_1}, \mathbf{O})$ and $\mathbf{Q}_n^* = \mathbf{I}_n - \mathbf{J}_{m_1,m_2}' \mathbf{J}_{m_1,m_2}$ (see Sect. 5.4 for a further extension).

If there are several seasonal frequencies, we need more complicated filtering procedures.

3.2 On Statistical Foundation

3.2.1 An Asymptotic Theory

At first glance, the SIML filtering procedure based on (3.1) and (3.2) may be regarded as an *ad-hoc* statistical procedure without any mathematical foundation. However, it has a rather solid statistical foundation.

Let $\theta_{jk} = \frac{2\pi}{2n+1}(j - \frac{1}{2})(k - \frac{1}{2})$, $p_{jk}^{(n)} = \frac{1}{\sqrt{2n+1}}(e^{i\theta_{jk}} + e^{-i\theta_{jk}})$ and for $\mathbf{Y}_n = (\mathbf{y}_i')$.
We write $\mathbf{z}_k^{(n)}$ $(k = 1, \ldots, n)$ as

$$\mathbf{z}_k^{(n)}(\lambda_k^{(n)}) = \sum_{j=1}^{n} p_{jk}^{(n)} \mathbf{r}_j, \quad \mathbf{r}_j = \mathbf{y}_j - \mathbf{y}_{j-1}, \tag{3.11}$$

which is a (real-valued) Fourier-type transformation and $\mathbf{y}_0$ is fixed.

Then, we find that $\mathbf{z}_k^{(n)}(\lambda_k^{(n)})$ $(k = 1, \ldots, n)$ are the (real-valued) Fourier transform of data at the frequency $\lambda_k^{(n)}$ $(= (k - 1/2)/(2n + 1))$, which is a (real-part of) estimate of the orthogonal incremental process $\mathbf{z}(\lambda)$ $(0 \leq \lambda \leq 1/2)$, which is a continuous stochastic process.

We shall utilize an asymptotic theory for the stationary linear processes because the time series model defined by (2.9)–(2.13) can be regarded as a special case. Let

$$\mathbf{r}_i = \Delta\mathbf{y}_i = \boldsymbol{\mu} + \sum_{j=0}^{\infty} \mathbf{C}_j \mathbf{u}_{i-j} , \tag{3.12}$$

where $\boldsymbol{\mu}$ is a constant vector (we assume $\boldsymbol{\mu} = \mathbf{0}$ for simplicity in this section), $\mathbf{C}_j$ are $p \times p$ matrices, and $\mathbf{u}_i$ are a sequence of mutually independent random variables with $E[\mathbf{u}_i] = \mathbf{0}$, $E[\mathbf{u}_i\mathbf{u}_i'] = \boldsymbol{\Sigma}_u$ (> 0). The errors-in-variables model in Sect. 2.2.2 implies that the $p \times p$ matrices $\mathbf{C}_j$ satisfy $\|\mathbf{C}_j\| = O(\rho^j)$ for $0 \leq \rho < 1$ and $\sum_{h=0}^{\infty} \|\mathbf{C}_j\| < \infty$.

We then summarize the useful result of the consistency and asymptotic normality on $\mathbf{z}_k^{(n)}(\lambda_k^{(n)})$. Although it could be regarded as a direct extension of Theorem 8.4.3 of Anderson (1971) for discrete and (ergodic) stationary time series[1], it is useful to present the following form. The proof is given in the Appendix of this chapter.

---

[1] See Chap. 7 of Durrett (1991) for the exposition of ergodicity and stationary time series, for instance.

Proposition 3.1 *Let r_j $(j = 1, \ldots, n)$ be an ergodic stationary stochastic process given by (3.12) with $\mu = 0$, $\|C_j\| = O(\rho^j)$ (for $0 \le \rho < 1$), $\sum_{h=0}^{\infty} \|C_j\| < \infty$ and the fourth-order moments of each element of u_i are finite.*

Let also $z_k^{(n)}(\lambda_k^{(n)}) = \sum_{j=1}^{n} p_{jk}^{(n)} r_j$ and r_j be an ergodic stationary sequence with $E[r_j] = 0$, $\Gamma(h) = E(r_j r'_{j-h})$ for any $h = 0, \pm 1, \pm 2, \ldots$, and the (symmetrized real-valued) spectral density matrix

$$f_{SR}(\lambda) = \sum_{h=-\infty}^{\infty} \cos(2\pi h\lambda)\Gamma(h) \ , \ 0 \le \lambda \le \frac{1}{2} \tag{3.13}$$

is the positive definite and bounded (real-valued and symmetrized) spectral matrix. Assume that $\lambda_k^{(n)} \to s$, $\lambda_{k'}^{(n)} \to t$ as $n \to \infty$ for $0 < s < t < \frac{1}{2}$. Then, as $n \longrightarrow \infty$

$$\begin{bmatrix} z_k^{(n)}(\lambda_k^{(n)}) \\ z_{k'}^{(n)}(\lambda_{k'}^{(n)}) \end{bmatrix} \xrightarrow{w} N_{2p}\left[0, \begin{bmatrix} f_{SR}(s) & 0 \\ 0 & f_{SR}(t) \end{bmatrix}\right]. \tag{3.14}$$

This proposition covers the general model with (2.9)–(2.13) with the moment conditions because Δy_i are stationary. As the asymptotic variance-covariance matrix of the orthogonal random vectors $z_k^{(n)}(\lambda_k^{(n)})$ is the (symmetrized real) spectral density matrix, it can be estimated consistently. When we have noise terms as in Sect. 3.2.2, it is not possible to estimate the (long-run) variance-covariance matrix Σ_x of trend-cycle component simply by using the differenced time series $r_j(= \Delta y_j) = y_j - y_{j-1}$. It is because

$$E[r_j r'_j] = E[\Delta x_j \Delta x'_j] + E[\Delta s_j \Delta s'_j] + E[\Delta v_j \Delta v'_j]. \tag{3.15}$$

The SIML estimator of $f_{SR}(0)$ $(= f_{\Delta x}(0)) = \Sigma_x$ in the general case can be defined by

$$G_m = \frac{1}{m_n} \sum_{k=1}^{m} (z_k^{(n)}(\lambda_k^{(n)}))(z_k^{(n)}(\lambda_k^{(n)}))'. \tag{3.16}$$

Then we summarize the basic property of the SIML estimation of Σ_x. The following proposition is the same of the first part of Theorem A.1 in the Appendix of Chap. 5 in the general case.

Proposition 3.2 *Assume that the fourth-order moments of each element of $v_i^{(x)}$, v_i, and s_i $(i = 1, \ldots, n)$ in (2.9)–(2.13) are bounded. We set $m_n = [n^\alpha]$ $(0 < \alpha < 1)$.*

Then, as $n \longrightarrow \infty$

$$G_m \xrightarrow{p} \Sigma_x. \tag{3.17}$$

The above result has some implication on the use of the frequency domain analysis of (non-stationary) multiple time series. For $0 \le \lambda \le \frac{1}{2}$, let the symmetrized (real-valued) spectral density matrix of $\Delta \mathbf{y}_j$ $(j = 1, \dots, n)$ be $f_{SR,\Delta y}(\lambda) = (1/2)[f_{\Delta y}(\lambda) + \bar{f}_{\Delta y}(\lambda)]$, where $\bar{f}(\,\cdot\,)$ is the complex conjugate of f and we have the initial condition $\mathbf{y}_0$.

Similarly, we find that $f_{SR,\Delta x}(\lambda) = (1/2)[f_{\Delta x}(\lambda) + \bar{f}_{\Delta x}(\lambda)]$,
$f_{SR,\Delta s}(\lambda) = (1/2)(1 - e^{2\pi i \lambda})[f_s(\lambda) + \bar{f}_s(\lambda)](1 - e^{-2\pi i \lambda})$, and
$f_{SR,\Delta v}(\lambda) = (1/2)(1 - e^{2\pi i \lambda})[f_v(\lambda) + \bar{f}_v(\lambda)](1 - e^{-2\pi i \lambda})$.

They are the consequence of (2.14), (2.15), and (2.16) for the symmetrized spectral (real-valued) density matrices.

From this interpretation of the symmetrized spectral densities, it may be easy to find that $\mathbf{G}_m$ is a consistent estimator of the long-run variance-covariance matrix, which is the spectral density matrix at the zero frequency $f_{SR,\Delta x}(0)$ when we use the $\mathbf{Z}_n^*$- transformation of data.

3.2.2 A Frequency Interpretation of SIML Filtering

In the traditional statistical time series analysis for a stationary discrete (vector) process with the (complex-valued) spectral distribution F, it has a representation with right-continuous (complex-valued) orthogonal increments in the frequency domain (see Doob 1953; Brockwell and Davis 1990 for the details). Chapter 7 of Anderson (1971) is informative because of its discussion on real-valued representations although it has only univariate cases. The real-valued multivariate orthogonal processes and the spectral density matrix play important roles in our formulation.

For $\lambda_k^{(n)} = (k - 1/2)/(2n + 1)$ $(k = 1, \dots, n)$, we rewrite (3.11) as

$$\mathbf{z}_k^{(n)}(\lambda_k^{(n)}) = \sum_{j=1}^{n} \mathbf{r}_j \frac{2}{\sqrt{2n + 1}} \cos\left[2\pi \lambda_k^{(n)} \left(j - \frac{1}{2}\right)\right] \quad (k = 1, \dots, n). \quad (3.18)$$

Then, by using the inversion transformation with $\mathbf{P}_n$, we can represent the original time series as

$$\mathbf{r}_s^{(n)} = \sum_{k=1}^{n} p_{sk} \, \mathbf{z}_k^{(n)}(\lambda_k^{(n)}) (s = 1, \dots, n). \quad (3.19)$$

It is another representation of $\mathbf{R}_n = (\mathbf{r}_i^{(n)'}) = \mathbf{C}_n^{-1}\hat{\mathbf{X}}_n(Q)$ in (3.1) when $\mathbf{Q}_n = \mathbf{I}_n$. For any s $(s = 1, \dots, n)$, $\mathbf{r}_s^{(n)}$ can be recovered as the weighted sum of orthogonal processes $\mathbf{z}_n(\lambda_k^{(n)})$ at frequency $\lambda_k^{(n)}$ $(k = 1, \dots, n)$. We then, by using $\mathbf{Y}_n = \mathbf{C}_n\mathbf{R}_n$, recover the non-stationary process $\mathbf{y}_t^{(n)}$ $(t = 1, \dots, n)$ given the initial condition $\mathbf{y}_0$ as

$$\mathbf{y}_t^{(n)} = \mathbf{y}_0 + \sum_{s=1}^{t} \mathbf{r}_s^{(n)}. \quad (3.20)$$

Let

$$
\alpha_n(\lambda_m^{(n)}, j - \frac{1}{2}) = \frac{1}{n} \sum_{k=1}^{m} \left[2 \cos 2\pi \lambda_k^{(n)} \left(j - \frac{1}{2} \right) \right].
\tag{3.21}
$$

Then, when $\lambda_m^{(n)} (= (m - 1/2)/(2n + 1)) \to \lambda$ as $n \to \infty$ $(0 < \lambda < \frac{1}{2})$, using the relation $\sum_{k=1}^{m} \cos 2\pi \frac{k-1/2}{2n+1}(j - 1/2)h = (1/2)[\sin 2\pi mh(j - 1/2)/(2n + 1)]/[\sin \pi h(j - 1/2)/(2n + 1)]$ (Lemma 5.1 of Kunitomo et al. 2018), we find

$$
\alpha_n \left(\lambda_m^{(n)}, j - \frac{1}{2} \right) \to \alpha(\lambda, j - \frac{1}{2}) = \frac{2 \sin 2\pi \lambda(j - \frac{1}{2})}{\pi(j - \frac{1}{2})}.
$$

If we set the uncorrelated stochastic process of uncorrelated increments with continuous parameter λ $(0 \leq \lambda \leq \frac{1}{2})$ as $A_n(\lambda) = \sum_{j=1}^{n} \alpha(\lambda, j - \frac{1}{2}) \mathbf{r}_j^{(n)}$, then we find

$$
\int_0^{\frac{1}{2}} \cos \left[2\pi \lambda \left(s - \frac{1}{2} \right) \right] dA_n(\lambda) = \mathbf{r}_s^{(n)} \quad (s = 1, \ldots, n).
\tag{3.22}
$$

This corresponds to the continuous representation of a discrete (real-valued) stationary time series in the frequency domain (see Chap. 7 of Anderson (1971)). If we write the limit of $\mathbf{A}(\lambda) = \lim_{n \to \infty} \mathbf{A}_n(\lambda)$ (assuming it exists), the (real-valued) spectral distribution matrix F_{RS} for any $0 \leq \lambda_1 < \lambda_2 \leq 1/2$ can be defined as

$$
F_{RS}(\lambda_2 - \lambda_1) = \mathbf{E}[(\mathbf{A}(\lambda_2 - \lambda_1)\mathbf{A}(\lambda_2 - \lambda_1)'] = \int_{\lambda_1}^{\lambda_2} f_{RS}(\lambda) d\lambda
\tag{3.23}
$$

if F_{RS} is absolutely continuous and the matrix-valued density process $f_{RS}(\lambda)$ $(0 \leq \lambda_1 < \lambda_2 \leq 1/2)$ exists.

From (3.5), we set $\hat{\mathbf{R}}_n(m) = (\hat{\mathbf{r}}_i^{(m,n)'}) = \mathbf{C}_n^{-1} \hat{\mathbf{X}}_n^{(m)}$ and $\hat{\mathbf{r}}_i^{(m,n)}$ are $p \times 1$ vectors for $i = 1, \ldots, n$. If we write

$$
\hat{\mathbf{r}}_s^{(m,n)} = \sum_{k=1}^{m} p_{sk} \mathbf{z}_k^{(n)}(\lambda_k^{(n)}) \quad (s = 1, \ldots, m; 0 < m < n),
\tag{3.24}
$$

it is the trend-cycle estimate of the SIML filtering value for $\mathbf{r}_s^{(m,n)}$, which is the corresponding element of $\mathbf{R}_n(m)$ $(= \mathbf{C}_n \mathbf{X}_n^{(m)})$. It corresponds to

$$
\mathbf{r}_s^{(m,n)} = \sum_{k=1}^{m} p_{sk} \mathbf{z}_k^{(n)*}(\lambda_k^{(n)}) \quad (s = 1, \ldots, m; 0 < m < n),
\tag{3.25}
$$

where $\mathbf{z}_n^*(\lambda_k^{(n)})$ are constructed from the $n \times p$ hidden states matrix $\mathbf{X}_n$ instead of the observed $n \times p$ matrix data $\mathbf{Y}_n$. Hence (3.25) is the same as the element of $\mathbf{C}_n^{-1}\hat{\mathbf{X}}_n^{(m)}$ in (3.4), and for $\lambda_m^{(n)} = m/(2n)$ in the frequency domain it is a discrete version of

$$\hat{\mathbf{r}}_s^{(n)}(\lambda_m^{(n)}) = \int_0^{\lambda_m^{(n)}} \cos\left[2\pi\lambda\left(s - \frac{1}{2}\right)\right] dA_n(\lambda) . \tag{3.26}$$

Then, (3.26) is the same as the element of $\mathbf{C}_n^{-1}\mathbf{X}_n^{(m)}$ in (3.4), and it has the corresponding (continuous) version in the frequency domain.

Similarly, $\hat{\mathbf{r}}_s^{(m_1,m_2,n)} = \sum_{k=m_1+1}^{m_1+m_2} p_{sk}\mathbf{z}_n(\lambda_k^{(n)}) = \hat{\mathbf{r}}_s^{(m_2,n)} - \hat{\mathbf{r}}_s^{(m_1,n)}$ $(s = 1, \ldots, m;$ $0 < m_1 < m_2 < n)$ can be regarded as a discrete version of

$$\hat{\mathbf{r}}_s^{(n)}(\lambda_{m_1}^{(n)}, \lambda_{m_2}^{(n)}) = \int_{\lambda_{m_1}^{(n)}}^{\lambda_{m_2}^{(n)}} \cos\left[2\pi\lambda\left(s - \frac{1}{2}\right)\right] dA_n(\lambda) . \tag{3.27}$$

From our interpretation of the SIML filtering we find a representation of (discrete time and real-valued) stationary processes and orthogonal incremental stochastic processes. They are closely related to our method of data analysis for non-stationary vector time series.

3.2.3 On Multivariate Filtering Methods

There may be a natural question whether any multivariate method of detrending or seasonal adjustment is necessary because the filtered series can depend on the particular choice of co-variables. This may be true if we use some time domain models such as multivariate ARMA models or other multivariate transformations. An application of Kalman-filtering approach may be an example because it has a multivariate state equation. Also the application of univariate ARIMA models with X-12-ARIMA in the seasonal adjustment filtering has another problem because the filtering result depends on ARIMA for each series and there is no guarantee to keep the original relation among variables at each frequency of our interest. Some filters such as the HP filter use the minimization of univariate criterion function with restriction. Then the filtered result depends on the choice of criteria for each series and there is no guarantee to preserve the original relation among variables at each frequency.

Because our method solely depends on the frequency decomposition, it may be free from these problems. Hence, it should be a robust filtering method with this respect and it may be appropriate for analyzing non-stationary multivariate time economic series.

3.3   Some Simulation and the Choice of Frequencies

3.3.1   A Guide of Choosing Frequencies

When we are interested in filtering a non-stationary time series with trend-cycle, seasonal, and measurement error components, we need to choose the parameter m (or m_1, m_2). We give a guide to set m in the trend-cycle filtering case for practical purpose. From the discussion of Sect. 3.2.2, the orthogonal process $\mathbf{z}_n(\lambda_k^{(n)})$ corresponds to the frequency $\lambda_k^{(n)} = (k - \frac{1}{2})/(2n + 1)$ $(k = 1, \ldots, n)$. When we are interested in the trend-cycle component, we may only use $\lambda_k^{(n)} = (k - \frac{1}{2})/(2n + 1)$ $(k = 1, \ldots, m)$ and then the maximum frequency is approximately

$$\lambda_{max}^{(n)} = \frac{m}{2n}. \tag{3.28}$$

For instance, when we have monthly data over 20 years as an example, we have $n = 240$ and $s = 12$. Although we have seasonal frequencies, we want to find trend-cycle components as business cycles more than 1.5 year, say. Then, an appropriate maximum frequency would be $\lambda_{max}^{(n)} = 1.5/24$ and then we could take $m^* = 480 \times (1.5/24) = 30$.

As another example if we have quarterly data over 30 years, we have $n = 120$ and $s = 4$. Although we have seasonal frequency, we want to find trend-cycle components as business cycles more than 1.5 year, say. Then an appropriate maximum frequency would be $\lambda_{max}^{(n)} = 1.5/8$ and then we could take $m^* = 240 \times (1.5/8) = 45$.

If we were interested in the trend-cycle component of the non-stationary time series, these choices might be reasonable candidates.

As we shall see in the next subsection, this point could be checked by simulation for prediction.

3.3.2   Some Simulation

When we have estimates of the state variables $\mathbf{x}_i^{(m)}$ $(i = 1, \ldots, n)$, the estimates of error components are $\hat{\mathbf{v}}_i^{(m)} = \mathbf{y}_i - \hat{\mathbf{x}}_i^{(m)}$ $(i = 1, \ldots, n)$. Then, an estimated MSE of the h-step ahead prediction error based on the SIML smoothing or filtering is given by

$$\mathrm{PMSE}_n(h) = \mathbf{E}[(\mathbf{y}_{n+h} - \hat{\mathbf{y}}_{n+h|n})(\mathbf{y}_{n+h} - \hat{\mathbf{y}}_{n+h|n})' | \mathcal{F}_n], \tag{3.29}$$

where $\hat{\mathbf{y}}_{n+h|n}$ is the predictor given $\mathcal{F}_n$ and $\mathcal{F}_n$ is the σ-field (information) available at n.

In the filtering problem, we may use the last h periods in the sample to evaluate the goodness of fit. One may try to minimize the estimated h-step prediction MSE by choosing an appropriate m by using the last part of the sample. It may be reasonable to take $h = 2, 3$ as short-run prediction and $h = 4, 8$ for long-run prediction in our limited experiments.

Although we have done a number of simulations, we only show a result on the simple trend plus noise model when $p = 1$, $x_i = x_{i-1} + v_i^{(x)}$, and $y_i = x_i + v_i$ ($i = 1, \ldots, n$) with $v_i^{(x)} \sim N(0, \sigma_x^2)$ and $v_i \sim N(0, \sigma_v^2)$. The criterion function is the prediction MSE given by

$$\mathrm{PMSE}_n^*(h) = \frac{1}{h} \sum_{i=n-h+1}^{n} (y_i - \hat{x}_i^{(m)})^2 \, ,$$

where $\hat{x}_i^{(m)}$ is the estimate of $x_i^{(m)}$.

We present the minimum m as m^* based on the trend-cycle filtering as Tables 3.1, 3.2, and 3.3 by taking $h = 2, 4, 6, 8$ and $n = 80, 120, 200, 400$. In our simulations, as n increases, we have larger choice of m^* while m^* decreases as h increases. When we have a long horizon with h, it may be natural to use a small number of lower frequencies, and for a wide range of σ_x and σ_v.

Table 3.1 PMSE and choice of m ($\sigma_x = 0.3$, $\sigma_v = 0.05$)

n	80	120	200	400
$h = 2$	12	19	32	65
$h = 3$	8	12	21	42
$h = 4$	6	9	15	32
$h = 5$	5	7	13	26
$h = 6$	4	6	10	21
$h = 7$	4	5	9	18
$h = 8$	3	4	8	16

Table 3.2 PMSE and choice of m ($\sigma_x = 0.3$, $\sigma_v = 0.4$)

n	80	120	200	400
$h = 2$	13	19	33	66
$h = 3$	8	13	21	43
$h = 4$	6	9	16	32
$h = 5$	5	8	13	26
$h = 6$	4	6	11	21
$h = 7$	4	5	9	18
$h = 8$	3	5	8	16

Table 3.3 PMSE and choice of m ($\sigma_x = 0.3$, $\sigma_v = 1.0$)

n	80	120	200	400
$h = 2$	13	20	33	67
$h = 3$	9	13	22	44
$h = 4$	6	10	16	33
$h = 5$	5	8	13	26
$h = 6$	4	7	11	22
$h = 7$	4	6	9	19
$h = 8$	4	5	8	16

Fig. 3.1 An example of SIML and HP filters. (An illustrated comparison of the SIML and HP Filters by using consumption data.)

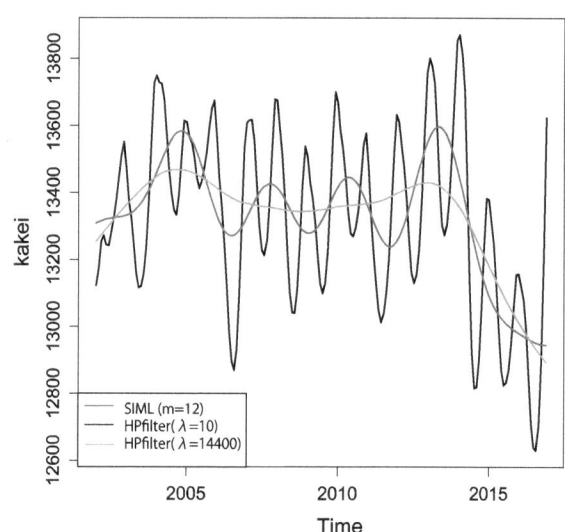

3.3.3 A Comparison of the SIML Filtering and HP Filtering

In order to characterize some property of the SIML filtering, we give an illustrative example. For this purpose, we take a monthly original consumption data in Chap. 2 (Kakei-Chosa series in Fig. 2.3) and show the result of the SIML filtering and the HP filtering with two parameter values (λ) as Fig. 3.1. (We used $hpfilter$ procedure in mFilter R-package for HP filter and took $m = 12$ for the SIML filter, which corresponds to frequencies over 2 year cycles since $\lambda_{max}^{(n)} = 1/24$.) An important point is that when we have strong seasonality in data, the SIML often gives reasonable estimates while the usefulness of HP filter depends on the the particular parameter value chosen for λ.

Further comparison would be interesting for practical purpose, but it differs from our main theme and it is beyond the scope of this book.

3.4 Applications

3.4.1 An Interpretation of Müller and Watson (2018)

As a related study in econometrics, Müller and Watson (2018) have proposed the
so-called long-run co-variability of macroeconomic time series. They have investigated many non-stationary time series and obtained some empirical findings. In this
subsection, an interpretation of their method will be given as measuring the relationships among long-run trend-cycle in our framework when $p = 2$. Moreover, we
obtain an important consequence from Proposition 3.2 in Sect. 3.2.1.

In this subsection, we consider the case of $p = 2$ in (2.9)–(2.13). Let 2×2
matrices $\boldsymbol{\Sigma}_x = (\sigma_{ij}^{(x)})$, we then define the (long-run) regression coefficient $\beta = [\sigma_{22}^{(x)}]^{-1}\sigma_{21}^{(x)}$ under the assumption that $\sigma_{22}^{(x)} > 0$. Let also $\boldsymbol{G}_m(0) = (g_{ij}^{(m)})$, and an
$n \times 2$ matrix

$$(\mathbf{w}_1, \mathbf{w}_2) = \mathbf{C}_n^{-1}(\mathbf{Y}_n - \mathbf{Y}_0) . \tag{3.30}$$

For estimating β, we define

$$\hat{\beta} = [g_{22}^{(m)}]^{-1}g_{21}^{(m)} = [\mathbf{w}_2'\mathbf{P}_n\mathbf{J}_m\mathbf{J}_m'\mathbf{P}_n\mathbf{w}_2]^{-1}[\mathbf{w}_2'\mathbf{P}_n\mathbf{J}_m\mathbf{J}_m'\mathbf{P}_n\mathbf{w}_1] . \tag{3.31}$$

This quantity can be interpreted as the least squares slope of the transformed vector
from $\mathbf{y}_{1n}$ on the transformed vector from $\mathbf{y}_{2n}$ for an $n \times 2$ matrix $\mathbf{Y}_n = (\mathbf{y}_{1n}, \mathbf{y}_{2n})$,
which is essentially the same as the one proposed by Muller and Watson (2018).[2]
Then, from Proposition 3.2 we obtain the following result, which is a special case of
the first part of Theorem 5.1 in Chap. 5.

Proposition 3.3 *In* (2.9)–(2.13) *with* $p = 2$, *we assume that the fourth-order
moments of each element of* $\mathbf{v}_i^{(x)}$, $\mathbf{v}_i$, *and* $\mathbf{v}_i^{(s)}$ ($i = 1, \ldots, n$) *are bounded, and* $\boldsymbol{\Sigma}_x$ *is
positive definite.*

(i) We fix an m, then $\hat{\beta}$ *is not consistent when* $n \to \infty$.
(ii) Set $m_n = [n^{\alpha}]$ *and* $0 < \alpha < 1$, *then as* $n \longrightarrow \infty$

$$\hat{\beta} - \beta \xrightarrow{p} \mathbf{0} . \tag{3.32}$$

A Simulation Example
To illustrate our arguments in Proposition 3.3, we performed a set of Monte Carlo
experiments under the simple situation with $p = 2$ (the replication was 10,00 in each
simulation). The model is given by

$$\mathbf{x}_i = \begin{pmatrix} x_{1,i} \\ x_{2,i} \end{pmatrix} = \begin{pmatrix} x_{1,i-1} \\ x_{2,i-1} \end{pmatrix} + \begin{pmatrix} v_{1,i}^{(x)} \\ v_{2,i}^{(x)} \end{pmatrix} , \tag{3.33}$$

---

[2] In their notation, m corresponds to q, which is fixed. They did use the (differenced) stationary
data, and thus, we could interpret that they calculated the linear regression from the filtered data
$\hat{\mathbf{X}}_n^* = \mathbf{P}_n'\mathbf{J}_m'\mathbf{J}_m\mathbf{P}_n\mathbf{C}_n^{-1}(\mathbf{Y}_n - \mathbf{Y}_0)$ as a modification of (3.1) in our notation.

Table 3.4 Simulations (direct LS)

n	200	1000	30000
True	1.2000	1.2000	1.2000
Mean	2.2016	2.2287	2.0918
SD	0.0930	0.2183	1.4901

$$\mathbf{y}_i = \begin{pmatrix} y_{1,i} \\ y_{2,i} \end{pmatrix} = \begin{pmatrix} x_{1,i} + \beta x_{2,i} \\ x_{2,i} \end{pmatrix} + \begin{pmatrix} v_{1,i} \\ v_{2,i} \end{pmatrix}, \tag{3.34}$$

where we have generated the normal errors $\mathbf{v}_i^{(x)'} = (v_{1,i}^{(x)}, v_{2,i}^{(x)})$, $\mathbf{v}_i' = (v_{1,i}, v_{2,i})$, $(i = 1, \ldots, n)$ with zero means, variances $\sigma_j^{(x)} = 0.75$ and $\sigma_j^{(v)} = 1.5$ $(j = 1, 2)$, zero covariances, and an initial value $\mathbf{x}_0$.

This is a two-dimensional I(1) process, which is not cointegrated, but close to a co-integrated process. We present the finite sample properties of the (naive) least squares (LS) estimator from the original data, the Müller-Watson (MW) estimator, and the SIML (SIML) estimator from the transformed data.

The simulation results in Tables 3.4, 3.5, and 3.6 are consistent with our arguments in Proposition 3.3. In the tables, Direct LS stands for the standard regression on $\mathbf{y}_{1n}$ on $\mathbf{y}_{2n}$, while MW and SIML stand for Muller-Watson and SIML, respectively. When we have two-dimensional time series and they are not co-integrated, the standard least square method is badly biased. The procedure proposed by Müller and Watson (2018) often gives reasonable results, but its variance does not decrease as the sample size increases. The SIML estimation method has reasonable finite sample properties as well as reasonable asymptotic properties and it is applicable to more general cases.

Table 3.5 Simulations (MW)

n	200	1000	30000
True	1.2000	1.2000	1.2000
Mean	1.1418	1.2129	1.1875
SD	0.5279	0.5301	0.5337

Table 3.6 Simulations (SIML)

n	200	1000	30000
True	1.2000	1.2000	1.2000
Mean	1.0828	1.1807	1.2020
SD	0.3278	0.1904	0.0766

3.4.2 Some Macroeconomic Data and Monthly Consumption Index

As illustrations for the empirical analysis, we have used our filtering method to analyze Japanese quarterly (real) consumption-GDP data as the first example, and three sets of monthly consumption data as the second example, which have been discussed in Chap. 2. Appendix B of this chapter contains all figures in this subsection.

As the first example, the ratio of real GDP and real consumption (quarterly original series) and their time series plots are given in Figs. 2.2 and 2.3. They show non-stationarity in their trend-cycle and seasonality as typical macroeconomic variables. We then calculate the transformed data using $\mathbf{P}_n$ (which is not $\mathbf{K}_n^*$, but a kind of real Fourier transform) as Fig. 3.2 from the original consumption data. In this case, the transformed series gives a wild up and down fluctuation, and trend-cycle and seasonality are sucked. To find seasonal components from the original series, we calculated the realized z_k ($k = 1, \ldots, n$) (Fig. 3.3) and the empirical cumulative distribution of z_k^2 ($k = 1, \ldots, n$) (Fig. 3.4), which roughly correspond to the normalized and real-valued sample spectral distribution function. Each component of $\mathbf{z}_k$ ($k = 1, \ldots, n$) is an orthogonal decomposition in the frequency domain. Because we have quarterly macro-data, we have a large up and down around ($\lambda_k^{(n)} =$) 0.25, which corresponds to the seasonal frequency at $s = 4$. The empirical spectral distribution has an abrupt change at this frequency. From these figures, we can judge that the real Fourier transformation based on $\mathbf{K}_n^*$ does give useful information.

As it has been a practice in time series data analysis to use seasonal differencing $\Delta_s \mathbf{y}_i = \mathbf{y}_i - \mathbf{y}_{i-s}$ ($s = 4$) in the Box-Jenkins method, we calculated the real Fourier transform based on $\mathbf{P}_n$ (Fig. 3.5) after seasonal differencing. Although the contribution of the resulting orthogonal process around the seasonal frequency could be

Fig. 3.2 Orthogonal process $\mathbf{z}_k$ for original series. (Without differencing $\mathbf{C}_n^{-1}$ we used the transformation $\mathbf{P}_n$ of data and then calculated $\mathbf{z}_k$ ($k = 1, \ldots, n$). Data are the quarterly real consumption between 1994Q1–2018Q2, published in 2018 by the Economic Social Research Institute (ESRI), Cabinet Office, Japan.)

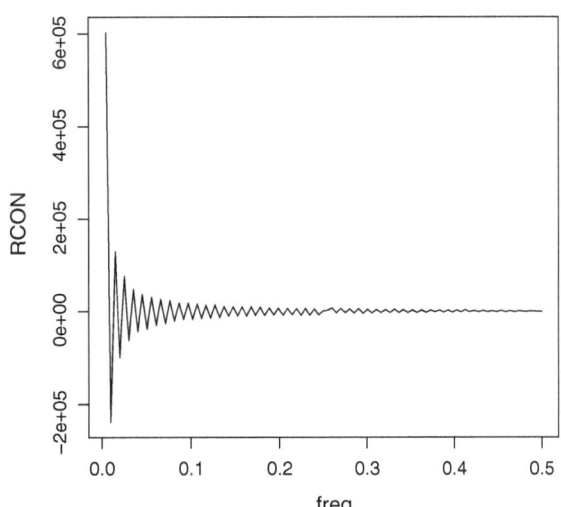

Fig. 3.3 Orthogonal process z_k for quarterly consumption. (We used the transformation K_n of data and then calculated z_k ($k = 1, \ldots, n$). Data are the quarterly real consumption between 1994Q1 and 2018Q2, published in 2018 by the Economic Social Research Institute (ESRI), Cabinet Office, Japan.)

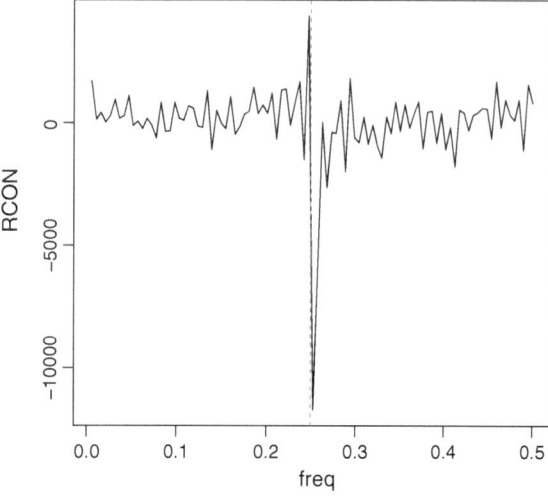

Fig. 3.4 Squared orthogonal process z_k^2 for quarterly consumption. (We calculated the cumulative sum of squares of the transformed orthogonal z_n^2 from the data.)

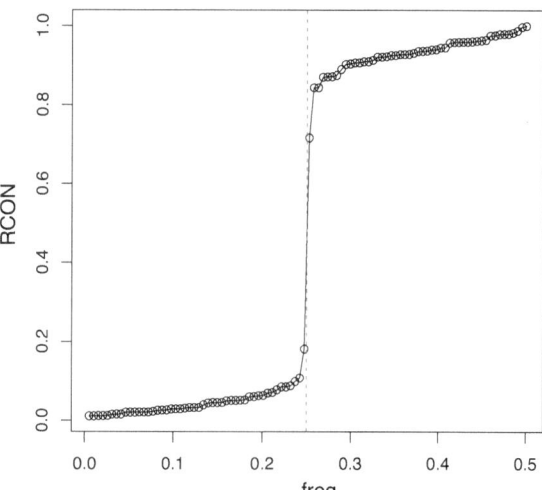

significant, there are some rather wild fluctuations at many other frequencies by using P_n-transformation (which is not K_n^*). Because we have some difficulty in interpreting the resulting time series, it may not be possible to justify the seasonal differencing procedure, and we recommend not to use this representation. In the following analysis, we simply use the differencing and then use the frequency domain analysis.

In Fig. 3.6, we have analyzed real Quarterly GDP. We show one example with $m = [n^{0.99}]$ and the deleted seasonal frequency is around 48–52 (48/196–52/196 in $[0, 1/2]$) and we delete extremely high-frequency part. We also deleted only five transformed data around the seasonal frequency and the main intention was to investigate the effect of seasonality with a narrow band. We have taken $\alpha = 0.99$ since

Fig. 3.5 Orthogonal process $\mathbf{z}_k$ for seasonally differenced consumption. (After seasonal differencing instead of C_n^{-1}, we used the transformation $\mathbf{P}_n$ on the differenced data and calculated the orthogonal elements.)

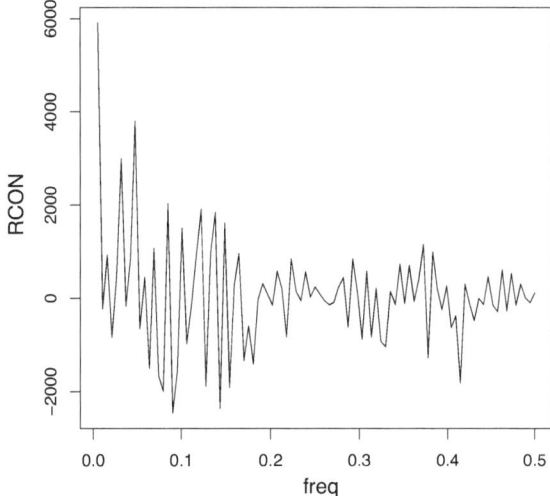

Fig. 3.6 Seasonally adjusted series. (The official seasonally adjusted (S.A.) series are constructed using X-12-ARIMA and we compared the SIML filtered value. Data are the quarterly real GDP between 1994Q1 and 2018Q2, published in 2018 by the Economic Social Research Institute (ESRI), Cabinet Office, Japan.)

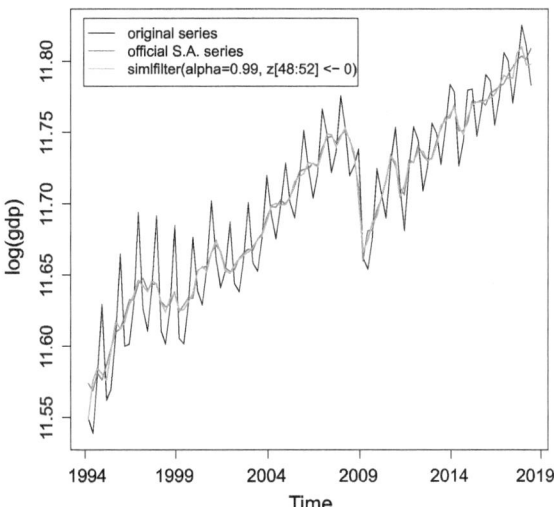

we wanted to remove some contribution of high frequency, but we could have used other choices and the results are not much different from Fig. 3.6 in most cases. We compared the filtered time series using our method and the official (published) seasonally adjusted time series. We found that the differences in two time series (i.e., the published time series from ESRI, Cabinet Office in Japan, and the SIML filtered time series) are rather small and they are often of negligible magnitude. Although our filtering procedure is simple, this empirical example suggests the usefulness of our method developed in this study.

As the second example, we have analyzed three consumption (monthly) time series and the quarterly consumption time series, which are mentioned in Chap. 2. This can be regarded as a typical empirical example when $p = 4$ and $s = 12$ with missing observations because the quarterly one series cannot give monthly information. As we have seen in Fig. 2.3, three monthly consumption series have similarities and some differences in their components. In our example, our goal is to construct the monthly consumption index, which is consistent with the observed quarterly consumption time series in trend-cycle component (see Chap. 6 for an alternative analysis of this data set). Because of non-stationary trend-cycle, seasonality, and measurement errors, it may not be obvious to construct a useful consumption index using existing statistical tools.

Let Y_i $(i = 1, \ldots, n)$ be the target (quarterly) time series and Z_{jt} $(j = 1, 2, 3; t = 3(i - 1) + l, l = 1, 2, 3)$ be the j-th monthly time series ($t = 0$ is the initial period and we fix the initial values Y_0, Z_{j0}). Then the criterion function is

$$\text{MSE}(m, m_1, m_2, m_3, w_1, w_2, w_3)$$

$$= \sum_{i=1}^{n} \left[\Delta \hat{Y}_i^{(T)} - \sum_{j=1}^{3} w_j \left(\sum_{l=1}^{3} \Delta Z_{j,3(i-1)+l}^{(T)} \right) \right]^2, \qquad (3.35)$$

where $\Delta \hat{Y}_i^{(T)} = \hat{Y}_i^{(T)} - \hat{Y}_{i-1}^{(T)}$, (the trend part of the estimated ΔY_i because we observe quarterly data on Y_i), $\Delta Z_{jt}^{(T)} = Z_{jt}^{(T)} - Z_{j,t-1}^{(T)}$ (the trend parts of ΔZ_{jt}), and w_j $(j = 1, 2, 3)$ are (unknown) weight coefficients and m, m_j $(j = 1, 2, 3)$ are the numbers of trend-cycle filtering. In the above formulation, we need to measure the prediction errors based on differenced data because we have non-stationary trend-cycle component.

Using the least squares method, we minimized the MSE criterion with respect to the underlying parameters. The estimated weights w_j $(j = 1, 2, 3)$ are 3.69, 5.19, and 1.64 (while the measurement units are different, but their magnitudes are comparable to the published quarterly consumption level at 2002Q1–2016Q4), which are statistically significant at 1%. We have chosen $m = 29, m_1 = 36, m_2 = 23$, and $m_3 = 33$. Although it may be possible to use other possibilities, but in our limited experiments, we found some improvements in prediction error over other cases with different combinations of m and m_j $(j = 1, 2, 3)$.

The black curves are the original series and the red curves are the estimated trend curves in Figs. 3.7 and 3.8 for two monthly series. By taking relatively large m_j $(j = 1, 2, 3)$, we can recover the cycle components of each series, which are crucial as the indicators of macro-business condition. In Fig. 3.9, the green curve shows the predicted state value calculated from the latest observed (quarterly) data plus the predicted monthly part based on the estimated parameters. As there is no monthly observation of quarterly published consumption, we draw their latest (quarterly) level with the black curve and the estimated SIML (filtered) values with the red

Fig. 3.7 Monthly
consumption series. (We
compared the original series
and the SIML filtered value.
Data are the monthly real
consumption between
2002M1–2016M12,
published in 2017 by the
Statistics Bureau, Japan.)

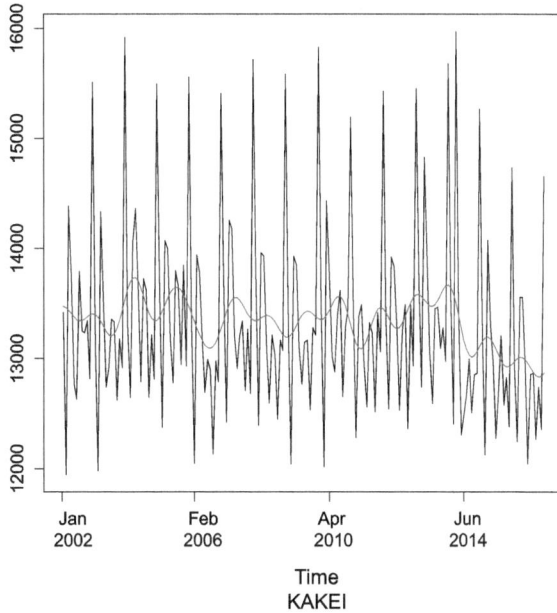

Fig. 3.8 Monthly
consumption series. (We
compared the original series
and the SIML filtered value.
Data are the monthly
consumptions between
2002M1 and 2016M12,
published in 2017 by
Ministry of Economy, Trade
and Industry (METI), Japan.)

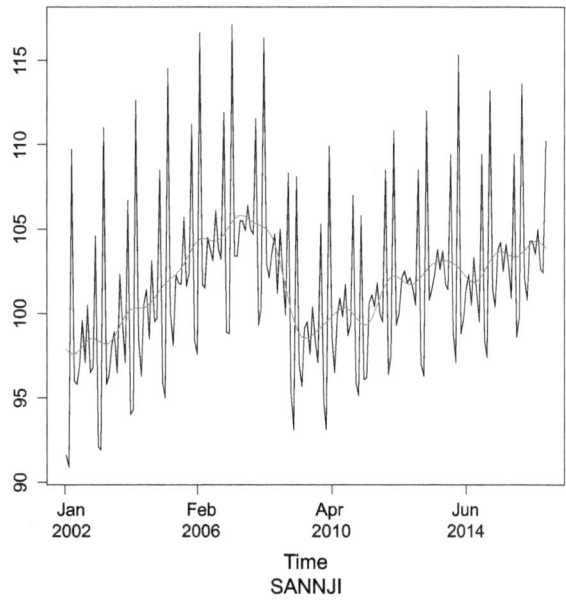

Fig. 3.9 Consumption series. (Data are the quarterly real consumption between 1994Q1–2018Q2, published in 2018 by the Economic Social Research Institute (ESRI), Cabinet Office, Japan.)

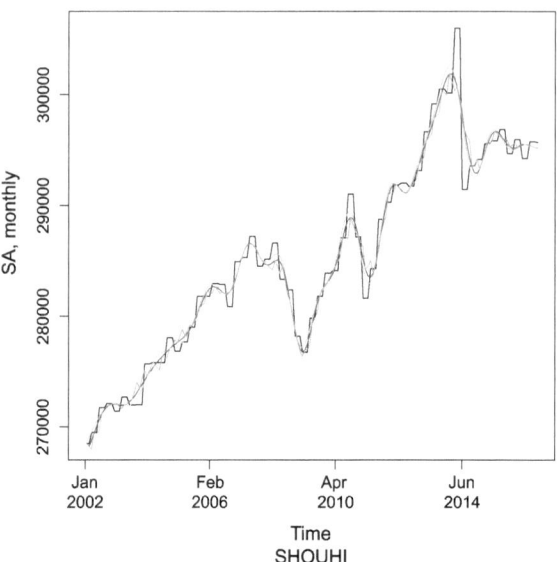

curve.[3] Overall, we found that while our procedure is much simpler than the X-12-ARIMA seasonal adjustment with Reg-ARIMA model, the estimated filtered series are consistent with the (published) quarterly series given the fact we have non-stationary trend-cycle, seasonal, and measurement errors components even when we do not have large samples. In Fig. 3.10, we have drawn the prediction errors in terms of the differenced value Y_i ($i = 1, \ldots, n$) based on our procedure. This figure illustrates the usefulness of the procedure because the macroeconomic time series are non-stationary with measurement errors.

The empirical data analyses in this subsection are presented for illustrations, but they suggest that the SIML filtering method is useful for real applications.

3.5   Some Remarks

When the observed non-stationary multivariate time series contains seasonal and noise components, it may be difficult to disentangle the effects of trend-cycle, seasonal, and measurement error components. In real (seasonally unadjusted) times series, we often observe non-stationary trend-cycle, seasonality, and measurement errors while the X-12-ARIMA program in official agencies uses the univariate Reg-ARIMA model to remove the seasonality from original time series. In this chapter, we investigate a new procedure to decompose time series into non-stationary trend-cycle

---

[3] One notable event was the introduction of consumption tax in April 2014 and a sharp deviation of trend. In the present study, we do not focus on this aspect and it will be in Sato and Kunitomo (2020).

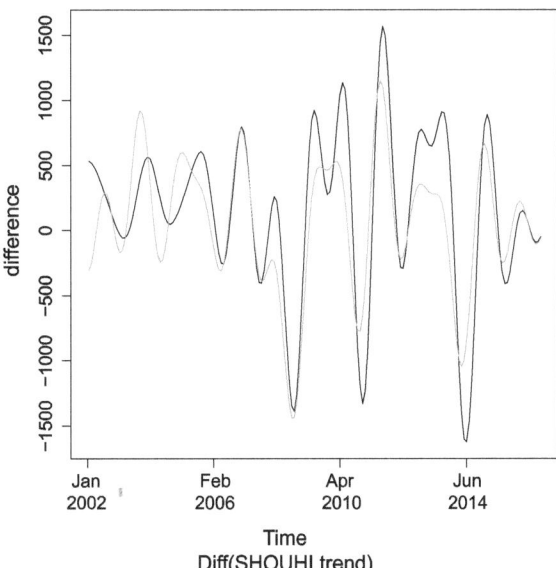

Fig. 3.10 Predicted Series. (The predicted and monthly consumption state variable are shown between 1994Q1 and 2018Q2.)

components and stationary seasonal and noise (or measurement errors) components. The resulting method for non-stationary multivariate series is simple and free from the underlying distributions of components. Hence, it is robust against possible misspecification in the non-stationary multivariate economic time series. An important conclusion is that it is useful to transform the observed time series using the $\mathbf{K}_n^*$-transformation and investigate the transformed $\mathbf{Z}_n$ series.

In empirical example in Sect. 3.4.2, we have illustrated our method to analyze quarterly and monthly macro consumption data in Japan. We presented one way to construct the monthly consumption index as the second example, which is consistent with the published or official (GDP-)consumption quarterly data. Although the problem is practically complicated, we have shown that our method gives a useful way for practical purpose. We will explore the problem further in Chap. 6.

There can be several further problems. Although it is easy to handle the $\mathbf{K}_n^*$-transformations of non-stationary multivariate time series and construct the transformed $\mathbf{Z}_n$-data, there is a problem of filtering when we have breaks and structural changes, for instance. It may not be easy to handle abrupt changes in multiple time series, such as the changes of consumption tax in Japan. We shall discuss these issues in Chap. 5.

Appendix A of this Chapter: Mathematical Derivations

We now present some details of derivations that we have omitted in this chapter. The proofs of Propositions 3.2 and 3.3 will be given in the Appendix of Chap. 5.

(i) **On** (3.6) **and** (3.9): When we take $m_1 = 0$ and $m_2 = m$, then (3.9) reduces (3.6). Hence, we show (3.9).

For $\theta_{jk} = \frac{2\pi}{2n+1}\left(j - \frac{1}{2}\right)\left(k - \frac{1}{2}\right)$ $(j, k = 1, \ldots, n)$, we use the relation that

$$\theta_{jk} + \theta_{j',k} = \frac{2\pi}{2n+1}(j + j' - 1)\left(k - \frac{1}{2}\right), \quad \theta_{jk} - \theta_{j',k} = \frac{2\pi}{2n+1}(j - j')\left(k - \frac{1}{2}\right).$$

For $h = 1, 2$, we then have

$$4 \sum_{k \in I_n^{(h)}} [\cos\theta_{jk} \cos\theta_{j',k}]$$

$$= \sum_{k \in I_n^{(h)}} [e^{i(\theta_{jk} + \theta_{j',k})} + e^{-i(\theta_{jk} + \theta_{j',+k})}]$$

$$+ \sum_{k \in I_n^{(h)}} [e^{i(\theta_{jk} - \theta_{j',k})} + e^{-i(\theta_{jk} - \theta_{j',k})}], \tag{3.36}$$

where $\mathbf{I}_n^{(1)} = [1, \ldots, m]$ and $\mathbf{I}_n^{(2)} = [m_1 + 1, \ldots, m_1 + m_2]$ are the index set for j and k.

For $\mathbf{I}_n^{(2)} = [m_1 + 1, \ldots, m_1 + m_2]$, by rewriting

$$\theta_{jk} + \theta_{j',k} = \left(m_1 - \frac{1}{2}\right)\frac{2\pi}{2n+1}(j + j' - 1) + \frac{2\pi}{2n+1}(j + j' - 1)(k - m_1),$$

and

$$\theta_{jk} - \theta_{j',k} = \left(m_1 - \frac{1}{2}\right)\frac{2\pi}{2n+1}(j - j') + \frac{2\pi}{2n+1}(j - j')(k - m_1).$$

The summation of the first term in (3.36) for $\mathbf{I}_n^{(2)}$ becomes

$$e^{i(m_1 + \frac{1}{2})\frac{2\pi}{2n+1}(j+j'-1)} \times \frac{1 - e^{i\frac{2\pi}{2n+1}(j+j'-1)m_2}}{1 - e^{i\frac{2\pi}{2n+1}(j+j'-1)}}$$

$$+ e^{-i(m_1 + \frac{1}{2})\frac{2\pi}{2n+1}(j+j'-1)} \times \frac{1 - e^{-i\frac{2\pi}{2n+1}(j+j'-1)m_2}}{1 - e^{-i\frac{2\pi}{2n+1}(j+j'-1)}}$$

$$= \frac{e^{i\frac{2\pi}{2n+1}m_1(j+j'-1)} - e^{i\frac{2\pi}{2n+1}(m_1+m_2)(j+j'-1)}}{e^{i\frac{2\pi}{2n+1}(-\frac{1}{2})(j+j'-1)} - e^{i\frac{2\pi}{2n+1}(\frac{1}{2})(j+j'-1)}}.$$

$$+\frac{e^{-i\frac{2\pi}{2n+1}m_1(j+j'-1)}-e^{-i\frac{2\pi}{2n+1}(m_1+m_2)(j+j'-1)}}{e^{-i\frac{2\pi}{2n+1}(-\frac{1}{2})(j+j'-1)}-e^{-i\frac{2\pi}{2n+1}(\frac{1}{2})(j+j'-1)}}$$

$$=\frac{\sin\frac{2(m_1+m_2)\pi}{2n+1}(j+j'-1)-\sin\frac{2(m_1)\pi}{2n+1}(j+j'-1)}{\sin\frac{\pi}{2n+1}(j+j'-1)}.$$

The summation of the second term in (3.36) is the same as the first term and then we use the coefficient terms as $[4/(2n+1)]\times(1/4)=1/(2n+1)$.

For the last two terms in (3.36), we use the same method of evaluation, but we need to evaluate each term when (i) $j=j'$ and (ii) $j\neq j'$, separately. Using similar calculations in (3.36) with the index set $\mathbf{I}_n^{(2)}$, when $j\neq j'$, the summation of the last two terms becomes

$$e^{i(m_1+\frac{1}{2})\frac{2\pi}{2n+1}(j-j')}\times\frac{1-e^{i\frac{2\pi}{2n+1}(j-j')m_2}}{1-e^{i\frac{2\pi}{2n+1}(j-j')}}$$

$$+e^{-i(m_1+\frac{1}{2})\frac{2\pi}{2n+1}(j-j')}\times\frac{1-e^{-i\frac{2\pi}{2n+1}(j-j')m_2}}{1-e^{-i\frac{2\pi}{2n+1}(j-j')}}.$$

$$=\frac{e^{i\frac{2\pi}{2n+1}m_1(j-j')}-e^{i\frac{2\pi}{2n+1}(m_1+m_2)(j-j')}}{e^{i\frac{2\pi}{2n+1}(-\frac{1}{2})(j-j')}-e^{i\frac{2\pi}{2n+1}(\frac{1}{2})(j-j')}}$$

$$+\frac{e^{-i\frac{2\pi}{2n+1}m_1(j-j')}-e^{-i\frac{2\pi}{2n+1}(m_1+m_2)(j-j')}}{e^{-i\frac{2\pi}{2n+1}(-\frac{1}{2})(j-j')}-e^{-i\frac{2\pi}{2n+1}(\frac{1}{2})(j-j')}}$$

$$=\frac{\sin\frac{2(m_1+m_2)\pi}{2n+1}(j-j')-\sin\frac{2(m_1)\pi}{2n+1}(j-j')}{\sin\frac{\pi}{2n+1}(j-j')}.$$

When $j=j'$, $\theta_{jk}-\theta_{j',k}=0$ and the summation of last two terms with the index set $\mathbf{I}_n^{(2)}$ becomes $2m_2$. Hence, we evaluate the corresponding results for $j-j'$ (there are two cases when (a) $j=j'$ and (b) $j\neq j'$), we obtain the results of (3.9), and then (3.6) by using $\mathbf{I}_n^{(1)}$ instead of $\mathbf{I}_n^{(2)}$ (and setting $m_1=0$ and $m_2=m$).

(ii) **The proof of Proposition 3.1**: Essentially, we apply the CLT (Central Limit Theorem, Theorem 8.4.3 of Anderson (1971) or Theorem 7.6 of Durrett (1991)) to the sequence of ergodic stationary (discrete) time series. We show the basic steps of our derivations and mention that the problem here is similar to those explained in Chaps. 7–9 of Anderson (1971) in detail.

First, we need to show that the resulting variance-covariance terms correspond to those of the limiting Gaussian random variables.

For this purpose, we need to evaluate

$$\mathbf{E}\left[\mathbf{z}_k^{(n)}(\lambda_k^{(n)})\mathbf{z}_{k'}^{(n)}(\lambda_{k'}^{(n)})'\right] = \left[\frac{1}{2n+1}\right]$$

$$\sum_{j,j'=1}^{n}(e^{i\theta_{jk}}+e^{-i\theta_{jk}})(e^{i\theta_{j'k'}}+e^{-i\theta_{j'k'}})\mathcal{E}[\mathbf{r}_j\mathbf{r}_{j'}'].\tag{3.37}$$

When $k \neq k'$, we find that the right-hand side terms are bounded by using the straight-forward calculations. We notice that the right-hand side consists of sums of four terms associated with

$$(e^{i\theta_{jk}}+e^{-i\theta_{jk}})(e^{i\theta_{j'k'}}+e^{-i\theta_{j'k'}})\tag{3.38}$$
$$= e^{i(\theta_{jk}+\theta_{j'k'})} + e^{-i(\theta_{jk}+\theta_{j'k'})} + e^{i(\theta_{jk}-\theta_{j'k'})} + e^{-i(\theta_{jk}-\theta_{j'k'})}$$
$$= (A) + (B) + (C) + (D) \ (, \text{ say}).$$

Then we find that the sums of each term associated with (A) and (B) in (3.38) are bounded, which becomes small when n is large. We write

$$\sum_{j,j'=1}^{n} e^{i(\theta_{jk}+\theta_{j'k'})}\mathcal{E}[\mathbf{r}_j\mathbf{r}_{j'}'] = \sum_{h=-(n-1)}^{n-1}\sum_{j'\in S(h)} e^{i(\theta_{h+j'_,k}+\theta_{j'k'})}\Gamma(h),\tag{3.39}$$

where $S(h) = \{1,2,\ldots,n-h\}$ for $h \geq 0$ and $S(h) = \{1-h,2-h,\ldots,n\}$ for $h < 0$. When $h \geq 0$ given h, the sum is given as

$$\sum_{j'=1}^{n-h} e^{i\frac{2\pi}{2n+1}(h+j'-\frac{1}{2})(k-\frac{1}{2})}e^{i\frac{2\pi}{2n+1}(j'-\frac{1}{2})(k'-\frac{1}{2})} \times \Gamma(h)$$

$$= \left[\sum_{j'=1}^{n-h} e^{i\frac{2\pi}{2n+1}(k+k'-1)(j'-1)}\right] e^{i\frac{2\pi}{2n+1}(h+1/2)(k-\frac{1}{2})+\frac{1}{2}(k'-\frac{1}{2})} \times \Gamma(h).$$

Because the sum in the parenthesis is

$$\frac{1-e^{i\frac{2\pi}{2n+1}(k+k'-1)(n-k)}}{1-e^{i\frac{2\pi}{2n+1}(k+k'-1)(n-k)}},$$

the terms associated with this factor are bounded when n is large, provided $k+k' > 1$. Similarly, when $h < 0$ given h, the sum can be written

$$\left[\sum_{j'=-h+1}^{n} e^{i\frac{2\pi}{2n+1}[(h+j'-1)(k+k'-1)]}\right] e^{i\frac{2\pi}{2n+1}+\frac{1}{2}(k-\frac{1}{2})-(h-\frac{1}{2})(k'-\frac{1}{2})} \times \Gamma(h).$$

It is also bounded when n is large. Then because these sums are finite and we have the condition $\sum_{h=-\infty}^{+\infty} \|\Gamma(h)\| < +\infty$, the sums with (A) and (B) in (3.38) become arbitrarily small when n is large. (The terms with (B) are the same as those with (A) except their signs in the exponential parts.)

Second, we need to show that the sums of each terms with (C) and (D) in (3.38) when $k = k'$ are dominant terms. We utilized the relation

$$
\sum_{j,j'=1}^{n} e^{i(\theta_{jk}-\theta_{j'k'})} \mathbf{E}[\mathbf{r}_j \mathbf{r}'_{j'}] = \sum_{h=-(n-1)}^{n-1} \sum_{j' \in S(h)} e^{i(\theta_{h+j',k}-\theta_{j'k'})} \times \Gamma(h). \qquad (3.40)
$$

When $h \geq 0$ given h, the second sum is given as

$$
\sum_{j'=1}^{n-h} e^{i\frac{2\pi}{2n+1}(j'-1)(k-k')} e^{i\frac{2\pi}{2n+1}[(h+\frac{1}{2})(k-\frac{1}{2})-\frac{1}{2}(k'-\frac{1}{2})]} \times \Gamma(h)
$$

and when $h < 0$ given h, the sum can be written

$$
\sum_{j'=-h+1}^{n} e^{i\frac{2\pi}{2n+1}[(h+j'-1)(k-k')+\frac{1}{2}(k-\frac{1}{2})-(h-\frac{1}{2})(k'-\frac{1}{2})]} \times \Gamma(h).
$$

The sums of the above terms are bounded when $k \neq k'$ by using the same argument as we have (3.40). (The terms with (D) are the same as those with (C) except their signs in the exponential parts.) On the other hand, when $k = k'$, $\sum_{j'=1}^{n-h} e^{i2\pi \frac{k-k'}{2n+1}(j'-1)} = n - h$. Then the dominant sums with (C) and (D) in (3.38) become

$$
\left[\frac{n}{2n+1}\right] \sum_{h=-(n-1)}^{n-1} \left[\cos 2\pi \frac{k-1/2}{2n+1} h\right] [\Gamma(h) + \Gamma(-h)]. \qquad (3.41)
$$

We note that under the assumption of stationarity of $\mathbf{r}_j$, it has been known that $\frac{1}{n}\sum_{j'=1}^{n} \mathbf{r}_{h+j'} \mathbf{r}'_{j'} \xrightarrow{P} \Gamma(h)$. (See Chap. 8 of Anderson (1971) and Brockwell and Davis (1990), for instance.)

Third, we consider the situation that $\lambda_k^{(n)} \to s$, $\lambda_{k'}^{(n)} \to t$ as $n \to \infty$ for $0 < s < t < \frac{1}{2}$. Since $\sum_{h=-\infty}^{+\infty} \|\Gamma(h)\| < +\infty$ and $\|\Gamma(h)\|$ is small as h is large, in the situation that $\lambda_k^{(n)} \to \lambda$ for $0 < \lambda < \frac{1}{2}$ as $n \to \infty$, we have (3.13). (We can take k such that $k/(2n) = \lambda + o(1/n^\epsilon)$ ($\epsilon > 0$) such that $\sum_{|h|>n^\epsilon} \|\Gamma(h)\|$ is arbitrary small.)

For the asymptotic normality, we set a sequence of random variables

$$
W_{k,k'}^{(n)} = (\boldsymbol{\alpha}'_1, \boldsymbol{\alpha}'_2) \begin{bmatrix} \mathbf{Z}'_n \mathbf{e}_k^{(n)} \\ \mathbf{Z}'_n \mathbf{e}_{k'}^{(n)} \end{bmatrix}, \quad k, k' = 1, \dots, n, \qquad (3.42)
$$

and $\mathbf{Z}_n = \mathbf{P}_n \mathbf{R}_n$, $\mathbf{R}_n = (\mathbf{r}'_j)$ ($n \times p$ matrix), where $\boldsymbol{\alpha}_i$ ($i = 1, 2$) are $p \times 1$ (non-zero) constant vectors and $e_k^{(n)} = (0, \ldots, 0, 1, 0 \ldots, 0)'$ ($k = 1, \ldots, n$) are $n \times 1$ unit vectors.

We can also construct a sequence of random variables $W_{s,t}^{(n)}$ by using $p_{s,j}^{(n)} = \sqrt{2/(2n+1)} \cos 2\pi s(j - 1/2)$ instead of $p_{jk}^{(n)}$ in $\mathbf{P}_n$.

Then $\mathcal{E}[|W_{k,k'}^{(n)} - W_{s,t}^{(n)}|^2] \xrightarrow{p} 0$ as $n \to \infty$ ($\lambda_k^{(n)} \to s$, $\lambda_{k'}^{(n)} \to t$ ($0 < s < t < \frac{1}{2}$). We use the fact that when $k \neq k'$ all terms in (3.38)–(3.40) are of $O(1)$. Then, $\mathbf{z}_k^{(n)}(\lambda_k^{(n)})$ and $\mathbf{z}_{k'}^{(n)}(\lambda_{k'}^{(n)})$ are asymptotically uncorrelated as $n \to \infty$.

Finally, we utilize the truncation argument used in the proof of Theorem 8.4.3 of Anderson (1971) and approximate (3.12) by the (finite order) moving average representation for a large $M > 0$

$$\mathbf{r}_i^{(n,M)} = \sum_{j=0}^{M} \mathbf{C}_j \mathbf{u}_{i-j} \tag{3.43}$$

and $\mathbf{z}_{n,M}(\lambda_k^{(n)}) = \sum_{j=1}^{n} p_{jk}^{(n)} \mathbf{r}_j^{(n,M)}$. By applying the standard method of proof in Theorem 8.4.3 of Anderson (1971), i.e., first we fix a $M > 0$ and we show the CLT for $\mathbf{z}_{n,M}(\lambda_k^{(n)})$ and then by taking $M \to \infty$ and we can show that the effects of M are negligible because $\sum_{s=-\infty}^{+\infty} \|\Gamma(s)\| < +\infty$.

We only mention to the key steps of the method used briefly because Anderson (1971, Chap. 8) gave the details of derivations for a slightly different problem. We use the decomposition $\mathbf{z}_{n,M}(\lambda_k^{(n)}) = \sum_{j=1}^{n} \sum_{s=0}^{M} p_{jk}^{(n)} \mathbf{C}_s \mathbf{u}_{j-s}$ into $\sum_{r=1}^{n-M} \sum_{s=0}^{M} [\cdot] + \sum_{j=1}^{M} \sum_{s=j}^{M} [\cdot] + \sum_{j=n-M+1}^{n} \sum_{s=0}^{j-(n-m)} [\cdot]$ and find that the effects of last two summations are stochastically negligible. Then, we evaluate the variance-covariance matrix of the first term as

$$\sum_{r=1}^{n-M} \left[\sum_{s=0}^{M} \sum_{s'=0}^{M} p_{r+s,k}^{(n)} p_{r+s',k}^{(n)} \mathbf{C}_s \mathbf{E}(\mathbf{u}_r \mathbf{u}'_r) \mathbf{C}'_{s'} \right]. \tag{3.44}$$

We find that after lengthy, but straightforward calculations, it is asymptotically equivalent to

$$\sum_{s=-M}^{M} \sum_{s'=-M}^{M} \frac{1}{2} \left[\exp(i2\pi \lambda_k^{(n)}(s - s')) + \exp(-i2\pi \lambda_k^{(n)}(s - s')) \right] \mathbf{C}_s \boldsymbol{\Sigma}_u \mathbf{C}'_{s'}$$

where $\lambda_k^{(n)} = (k - 1/2)/(2n + 1)$ and we set $\mathbf{C}_s = \mathbf{O}$ ($s < 0$) for convenience. (We need to evaluate the order of $\sum_{r=1}^{n-M} p_{r+s,k}^{(n)} p_{r+s',k}^{(n)}$ given s, s' as (3.37)–(3.40), for instance.) Then, by using the relation that $\Gamma(h) = \sum_{s=0}^{\infty} \mathbf{C}_s \boldsymbol{\Sigma}_u \mathbf{C}_{s-h}$, $\sum_{h=-\infty}^{\infty} \|\Gamma(h)\| < \infty$, and $\lim_{M\to\infty} \sum_{|h|>M} \|\Gamma(h)\| = 0$, it is asymptotically equivalent to the variance-covariance matrix in (3.14). By applying a central limit theorem (CLT) to the truncated terms of independent random variables $\sum_{r=1}^{n-m} [\sum_{s=0}^{M} p_{r+s,k}^{(n)} \mathbf{C}_s] \mathbf{u}_r$, we have the result as $n \to \infty$. (Q.E.D.)

Appendix B of this Chapter: Some Figures

We now present figures used in Sect. 3.4. As we have explained in Chap. 2, all data are official data published by ESRI (Economic and Social Research Institute), Cabinet Office of Japan, METI, and Statistics Bureau, Ministry of Internal Affairs and Communications. They are available from the government official website: e.stat in JAPAN.

References

Anderson TW (1971) The statistical analysis of time series. Wiley

Brillinger D, Hatanaka M (1969) An harmonic analysis of nonstationary multivariate economic processes. Econometrica 35:131–141

Brillinger D (1980) Time series: data analysis and theory, Expanded edn. Holden-Day

Brockwell P, Davis R (1990) Time series: theory and methods, 2nd edn. Wiley

Doob JL (1953) Stochastic processes. Wiley

Durrett R (1991) Probability: theory and examples. Duxbury Press

Kunitomo N, Sato S, Kurisu D (2018) Separating information maximum likelihood estimation for high frequency financial data. Springer

Kunitomo N, Sato S (2017) Trend, seasonality and economic time series: the non-stationary errors-in-variables models. SDS-4, MIMS, Meiji University. http://www.mims.meiji.ac.jp/publications/2017-ds

Müller U, Watson M (2018) Long-run covariability. Econometrica 86–3:775–804

Chapter 4
Comparing Estimation Methods of Non-stationary Errors-in-Variables Models

Abstract We investigate the estimation methods of the multivariate non-stationary errors-in-variables models when there are non-stationary trend components and the measurement errors or noise components. We compare the maximum likelihood (ML) estimation and the separating information maximum likelihood (SIML) estimation. The Gaussian likelihood function can have non-concave shape in some cases and the ML method works only when the Gaussianity of the non-stationary and stationary components holds with some restrictions in the parameter space. The SIML estimator has the asymptotic robust properties in more general situations. We study the finite sample and asymptotic properties of the ML and SIML methods for the non-stationary errors-in-variables models.

4.1 Introduction

The main purpose of this chapter is to compare the SIML estimation and the maximum likelihood (ML) estimation, which are two different methods to estimate the multivariate non-stationary errors-in-variables models when there are non-stationary trends and noise components. We investigate the finite and large sample properties of two estimation methods. An important finding is the fact that the Gaussian likelihood function may have non-concave shape in some case of the non-stationary errors-in-variables models although the ML method works well when the Gaussianity of non-stationary and stationary components holds with restrictions on the parameter space. We need to restrict the range of the signal-noise variance ratio and the measurement errors are not small. When the measurement error is small in a sense and/or there are co-integrated relations among trends with the rank of state variables being smaller than the dimension of observations, there could be a serious problem in the ML estimation under the assumption of Gaussian distributions. The SIML method, on the other hand, provides an alternative way to overcome the underlying difficulty in a non-parametric way. It has the asymptotic robust properties under general conditions of moments for consistency and asymptotic normality.

© The Author(s), under exclusive license to Springer Nature Singapore Pte Ltd. 2025   43
N. Kunitomo and S. Sato, *The SIML Filtering Method for Noisy Non-stationary Economic Time Series*, JSS Research Series in Statistics,
https://doi.org/10.1007/978-981-96-0882-9_4

4.2   On Estimation of Non-stationary Errors-in-Variables Models

4.2.1   Estimation Methods

In this chapter, let y_{ji} be the i−th observation of the j−th time series at i for $i = 1, \ldots, n$; $j = 1, \ldots, p$. We set $\mathbf{y}_i = (y_{1i}, \ldots, y_{pi})'$ be a $p \times 1$ vector and $\mathbf{Y}_n = (\mathbf{y}_i') \, (= (y_{ij}))$ be an $n \times p$ matrix of observations and denote $\mathbf{y}_0$ as the initial $p \times 1$ vector. We consider the situation when the underlying non-stationary trends $\mathbf{x}_i \, (= (x_{ji})) \, (i = 1, \ldots, n)$ are not necessarily the same as the observed time series and let $\mathbf{v}_i = (v_{1i}, \ldots, v_{pi})$ be the vector of noise components, which are independent of $\mathbf{x}_i$. Then we use the additive state space decomposition form as $\mathbf{y}_i = \mathbf{x}_i + \mathbf{v}_i \, (i = 1, \ldots, n)$, where $\mathbf{x}_i \, (i = 1, \ldots, n)$ are a sequence of non-stationary trend components satisfying $\Delta \mathbf{x}_i = (1 - \mathcal{L})\mathbf{x}_i = \mathbf{v}_i^{(x)}$ with $\mathcal{L}\mathbf{x}_i = \mathbf{x}_{i-1}$, $\Delta = 1 - \mathcal{L}$, $\mathbf{E}(\mathbf{v}_i^{(x)}) = \mathbf{0}$, $\mathbf{E}(\mathbf{v}_i^{(x)}\mathbf{v}_i^{(x)'}) = \mathbf{\Sigma}_x$, and $\mathbf{v}_i \, (i = 1, \ldots, n)$ are a sequence of (mutually) independent noise components with $\mathbf{E}(\mathbf{v}_i) = \mathbf{0}$, $\mathbf{E}(\mathbf{v}_i \mathbf{v}_i') = \mathbf{\Sigma}_v$. We assume that $\mathbf{v}_i^{(x)}$ and $\mathbf{v}_i$ are the sequence of i.i.d. random variables with $\mathbf{\Sigma}_v$ being non-negative definite and finite, and the random variables $\mathbf{v}_i^{(x)}$ and $\mathbf{v}_i$ are mutually independent.

To investigate statistical properties of alternative estimation methods, in this chapter we consider the situation when $\Delta \mathbf{x}_i$ and $\mathbf{v}_i \, (i = 1, \ldots, n)$ are mutually independent and each of the component vectors are independently, identically, and normally distributed as $N_p(\mathbf{0}, \mathbf{\Sigma}_x)$ and $N_p(\mathbf{0}, \mathbf{\Sigma}_v)$, respectively. We use an $n \times p$ matrix $\mathbf{Y}_n = (\mathbf{y}_i')$ and consider the distribution of $np \times 1$ random vector $(\mathbf{y}_1', \ldots, \mathbf{y}_n')'$. Given the initial condition $\mathbf{y}_0$, we have $\mathrm{vec}(\mathbf{Y}_n) \sim N_{n \times p}\left(\mathbf{1}_n \cdot \mathbf{y}_0', \mathbf{I}_n \otimes \mathbf{\Sigma}_v + \mathbf{C}_n \mathbf{C}_n' \otimes \mathbf{\Sigma}_x\right)$ as (2.3) in Chapt. 2.

Then, given the initial condition $\mathbf{y}_0$, the conditional maximum likelihood (ML) estimator can be defined as the solution of maximizing the conditional log-likelihood function except a constant as

$$L_n^* = \log |\mathbf{I}_n \otimes \mathbf{\Sigma}_v + \mathbf{C}_n \mathbf{C}_n' \otimes \mathbf{\Sigma}_x|^{-1/2}$$
$$-\frac{1}{2}[vec(\mathbf{Y}_n - \bar{\mathbf{Y}}_0)']'[\mathbf{I}_n \otimes \mathbf{\Sigma}_v + \mathbf{C}_n \mathbf{C}_n' \otimes \mathbf{\Sigma}_x]^{-1}[vec(\mathbf{Y}_n - \bar{\mathbf{Y}}_0)'], \quad (4.1)$$

where $\bar{\mathbf{Y}}_0 = \mathbf{1}_n \cdot \mathbf{y}_0'$. To develop the method of the SIML estimation, we use the K_n-transformation from $\mathbf{Y}_n$ to $\mathbf{Z}_n \, (= (\mathbf{z}_k'))$ by $\mathbf{Z}_n = \mathbf{K}_n \left(\mathbf{Y}_n - \bar{\mathbf{Y}}_0\right)$, $\mathbf{K}_n = \mathbf{P}_n \mathbf{C}_n^{-1}$. By using the spectral decomposition $\mathbf{C}_n^{-1} \mathbf{C}_n'^{-1} = \mathbf{P}_n \mathbf{D}_n \mathbf{P}_n'$ and $\mathbf{D}_n$ is a diagonal matrix with the k-th element $d_k = 2[1 - \cos(\pi(\frac{2k-1}{2n+1}))] \, (k = 1, \ldots, n)$. Then, the conditional likelihood function given the initial condition is proportional to

$$L_n = \sum_{k=1}^{n} \log |a_{kn}^* \mathbf{\Sigma}_v + \mathbf{\Sigma}_x|^{-1/2} - \frac{1}{2} \sum_{k=1}^{n} \mathbf{z}_k'[a_{kn}^* \mathbf{\Sigma}_v + \mathbf{\Sigma}_x]^{-1}\mathbf{z}_k, \quad (4.2)$$

where $a_{kn}^* \ (= d_k) \ = 4\sin^2\left[\frac{\pi}{2}\left(\frac{2k-1}{2n+1}\right)\right] \ (k = 1, \ldots, n)$.

We use the transformation on the non-stationary time series and use the random variables $\mathbf{z}_k \ (k = 1, \ldots, n)$, which follows $N_p(\mathbf{0}, \mathbf{\Sigma}_x + a_{kn}^* \mathbf{\Sigma}_v)$ and the coefficient a_{kn}^* is a dense sample of $4\sin^2(x)$ in $(0, \pi/2)$.[1]

The ML estimation of unknown parameters is defined as the maximization of (4.2) with respect to $\mathbf{\Sigma}_v$ and $\mathbf{\Sigma}_x$. Because of the coefficient $a_{kn}^* \ (k = 1, \ldots, n)$, the ML estimator is a complicated function of data and its computation is not a trivial task as we shall see in Sects. 4.3 and 4.4.

From the representation (4.2), it may be natural to use $\mathbf{z}_k \mathbf{z}_k'$ to estimate $a_{kn}^* \mathbf{\Sigma}_v + \mathbf{\Sigma}_x$ since it is the variance-covariance matrix of $\mathbf{z}_k$. We notice that $a_{kn}^* \to 0$ as $n \to \infty$ for a fixed k. When k is small, a_{kn}^* is small and we can expect that $k = k_n$ depending n is still small when n is large. However, $(1/m_n)\sum_{k=1}^{m_n} a_{kn}^*$ is not small if m_n is near to n, which suggests the condition $m_n/n \to 0$ as $n \to \infty$. The separating information maximum likelihood (SIML) estimator of $\hat{\mathbf{\Sigma}}_x$ in Sect. 2.2.1 is defined by

$$\hat{\mathbf{\Sigma}}_{x,SIML} = \frac{1}{m_n} \sum_{k=1}^{m_n} \mathbf{z}_k \mathbf{z}_k' . \tag{4.3}$$

This estimator of the variance-covariance of non-stationary trends uses the information on trends in the frequency domain, which corresponds to use only the trend parts without measurement errors from the time series observations. The interpretation of (4.3) from the frequency domain of non-stationary time series has been discussed in Sect. 4.2.

From our construction of the SIML estimation the essential features of estimation do not much depend on the presence of noise terms when the noise terms are stationary. This feature was the main reason for developing the SIML method by Kunitomo et al. (2018).

Let the quadratic variation of observed vectors $\mathbf{y}_i \ (i = 1, \ldots, n)$ be

$$\mathbf{QV}_y(1) = \sum_{i=1}^{n} (\mathbf{y}_i - \mathbf{y}_{i-1})(\mathbf{y}_i - \mathbf{y}_{i-1})', \tag{4.4}$$

where $\mathbf{y}_0$ is the initial vector.

If we denote $\hat{\mathbf{\Sigma}}_y = (1/n)\mathbf{QV}_y(1)$, we may require the condition $\hat{\mathbf{\Sigma}}_{x,SIML} \leq \hat{\mathbf{\Sigma}}_y$ in the sense of positive semi-definiteness because $\mathbf{y}_i$ includes noise terms in addition to the state variables $\mathbf{x}_i$.

For the SIML estimation $\hat{\mathbf{\Sigma}}_{x,SIML}$, the number of terms m_n should be dependent on n. In the representation of (4.3) we need the order requirement that $m_n = O([n^\alpha])$ and $0 < \alpha < 1$, which is the first property of the macro-SIML estimation. There is a trade-off between the bias and the variances of the SIML estimator for $\mathbf{\Sigma}_x$. Proposition 3.2

---

[1] We have used a_{kn}^*, which is slightly different a_k in Kunitomo et al. (2018) and the latter corresponds to $a_{kn} = n a_{kn}^*$. They have investigated the SIML estimation when the length of observation interval decreases in a fixed period as the high-frequency asymptotics.

of Chap. 3 has shown that when $m_n \to \infty$ as $n \to +\infty$ for the consistency we need the condition $0 < \alpha < 1.0$ while for the asymptotic normality we need the condition $0 < \alpha < 0.8$ under the assumption that the parameter matrices $\boldsymbol{\Sigma}_x > 0$ and $\boldsymbol{\Sigma}_v > 0$ are fixed. (See Theorem 4.2 in Sect. 4.5.) As an example, we may take $\alpha = 0.79$.

4.2.2 Non-stationary Time Series and Measurement Errors

In this subsection by using a simple example when $p = 1$ we illustrate the reason why the presence of noise term in the non-stationary time series, even if it is small, forces us to change the standard thinking on time series analysis.

In traditional econometric analysis of time series, the non-stationarity of economic times series has been often discussed, but there were not many discussions on the role of measurement errors. (See Engle and Granger (1987) and Johansen (1995) for instance.) The standard arguments for the integrated processes have been to use the Brownian functionals to describe the behaviors of integrated processes when the sample size is large. As a typical example, we generate a set of mutually independent variables $v_i^{(x)}$ $(i = 1, \ldots, n)$, which follows $\sqrt{12}(U(0, 1) - 0.5)$ and $U(0, 1)$ follows the uniform distribution. Then we generate x_i and y_i satisfying $x_i = x_{i-1} + v_i^{(x)}$, and $y_i = x_i + v_i$ $(i = 1, \ldots, n)$ by adding the measurement errors v_i, which are independent $N(0, 1)$ random variables. By replicating 1,000 times, the empirical distribution of $[\sum_{i=1}^{n} y_i]/[\sigma_x \sqrt{n^3}]$ (we have set $\sigma_x^2 = \mathbf{E}(v_i^{(x)2}) = 1$ and $x_0 = 0$) can be approximated[2] by the weak convergence to the Brownian functional

$$X(1) = \int\limits_0^1 B(s)ds , \qquad (4.5)$$

where $B(s)$ is the Brownian motion on $[0, 1]$ and it follows $N(0, 1/3)$. Figure 4.1 illustrates the fact that this approximation is reasonable even when $n = 30$. (The red curve shows $N(0, 1/3)$.) However, as illustrated in Chaps. 9 and 10 of Hayashi (2000), in order to make statistical inference on the unit roots and possible co-integrated vectors, we need an estimate of the variance and covariances of innovation terms of integrated processes such as σ_x^2, which are generally unknown. There should be a careful analysis on the role of innovations, but the easiest way to estimate the variance of innovation term of the integrated processes and to use the normalized sum of squared differences of observed time series as $\hat{\sigma}_y^2$ $(= (1/n) \sum_{i=2}^{n}(y_i - y_{i-1})^2)$ although there can be more sophisticated methods such as the maximum likelihood method in more general situations. As a simple illustration, we generate a set of observations y_i $(i = 1, \ldots, n)$ by adding the measurement errors v_i

---

[2] The sum $(1/n) \sum_{i=1}^{n} v_i$ converges to zero in probability and $(1/n) \sum_{i=1}^{n}[1/(\sigma_x \sqrt{n}) \sum_{j=1}^{i} v_j^{(x)}]$ converges weakly to a random variable, which is a functional of Brownian motions. See Chap. 7 of Durrett (1991), for instance.

Fig. 4.1 Normal approximation ($n = 30$)

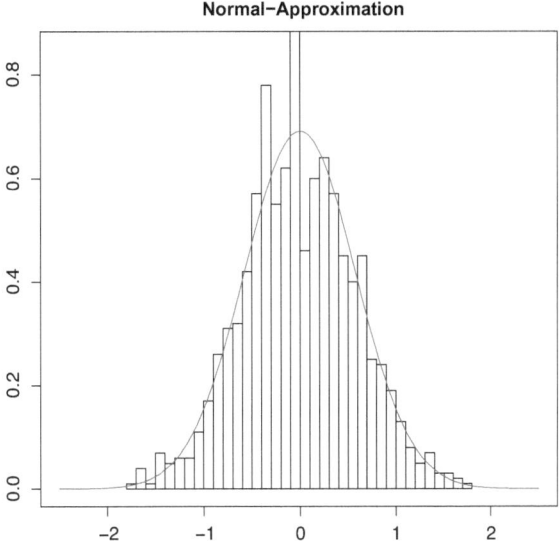

which are independent $N(0, 0.5)$ random variables to the I(1) process x_i, and thus $y_i = x_i + v_i$ ($i = 1, \ldots, n$). Figure 4.2 (Case 2-1) illustrates the estimated variance of innovation $v_i^{(x)}$ ($= x_i - x_{i-1}$) by this method, where the true parameter value (as red line) is 1. Although the variance of innovation of the integrated process is twice of the variance of measurement errors and the measurement errors is small in a sense, the standard estimated value has a significant bias and it is distributed around 2 when $n = 80$. This case corresponds to $c = 2 = \sigma_x^2/\sigma_v^2$. We also give Figs. 4.2, 4.3, 4.4, and 4.5 for Case 2-2 ($c = 1/2$), Case 2-3 ($c = 8$), and Case 2-4 ($c = 1/8$) for a comparison. Case 2-1 and Case 2-3 correspond to the small noise cases while Case 2-2 and Case 2-4 correspond to the large noise cases. In latter cases, the bias of the estimated variance σ_x^2 becomes large as we can expect.

The point here is the fact that even when we have small (Gaussian) noise we may have misleading estimation on the variance of system variables. These examples may illustrate the importance of our analysis of measurement errors in non-stationary time series. We usually do not have much information on the magnitude of c, which is the ratio of the signal variance over the noise variance, and the distribution of measurement errors in advance when we observe non-stationary data. Hence, it is important to use the statistical method which does not depend on the value of c and the distribution of noise for practical purpose.

Fig. 4.2 Estimated variance
($n = 80, c = 2$)

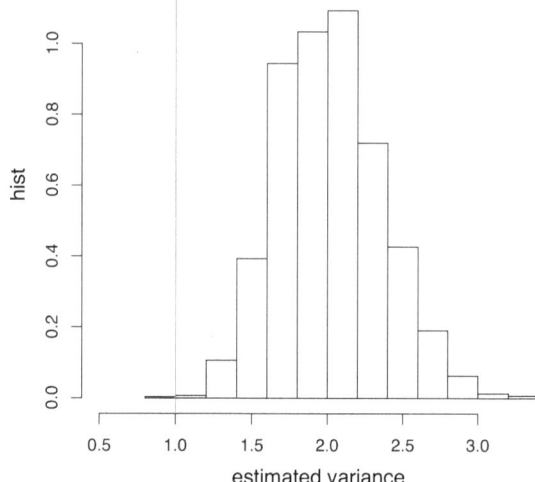

Fig. 4.3 Estimated variance
($n = 80, c = 1/2$)

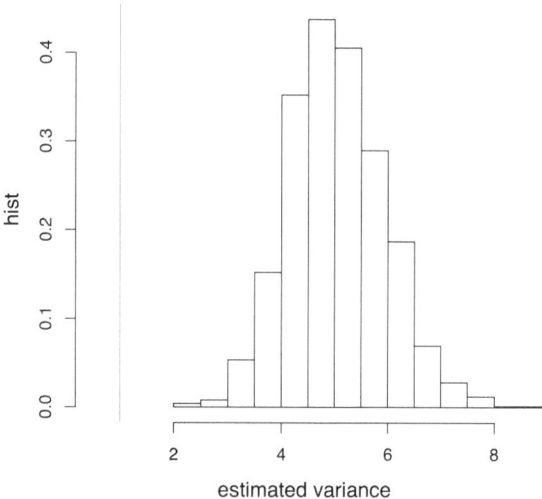

4.3   Simple Cases

4.3.1   An Illustrative Example

To see the main problem of our interest in a clear way, we consider the simplest case when $p = 1$. Let y_i be the i-th observation of time series for $i = 1, \ldots, n$ and $\mathbf{y}_n = (y_i)$ be an $n \times 1$ vector of observations (y_0 is the initial observation). We consider the situation when the underlying non-stationary trends x_i ($i = 1, \ldots, n$) satisfy

Fig. 4.4 Estimated variance ($n = 80, c = 8$)

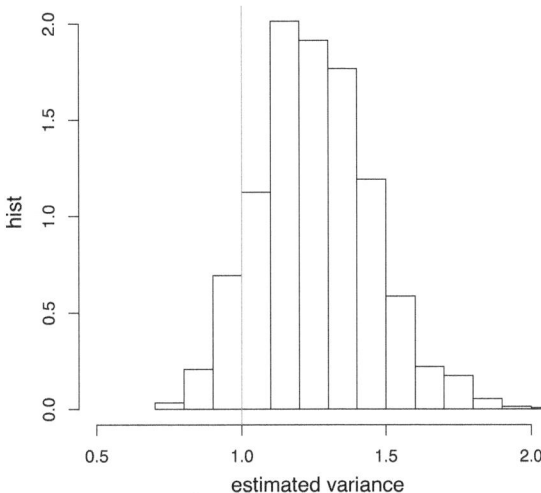

Fig. 4.5 Estimated variance ($n = 80, c = 1/8$)

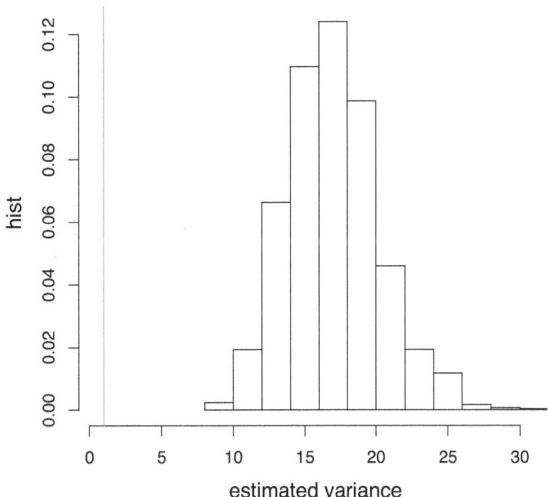

$$x_i = x_{i-1} + v_i^{(x)}, \tag{4.6}$$

where $v_i^{(x)}$ are the independent random variables followed by $N(0, \sigma_x^2)$ and x_0 is the initial variable. Let v_i (measurement error) be the sequence of i.i.d. random variables followed by $N(0, \sigma_v^2)$, which are independent of x_i ($i = 1, \ldots, n$). For the additive model

$$y_i = x_i + v_i \quad (i = 1, \ldots, n), \tag{4.7}$$

the log-likelihood function is proportional to

$$L_n = \sum_{k=1}^{n} \log |a_{kn}^* \sigma_v^2 + \sigma_x^2|^{-1/2} - \frac{1}{2} \sum_{k=1}^{n} \frac{z_k^2}{a_{kn}^* \sigma_v^2 + \sigma_x^2}, \qquad (4.8)$$

where

$$a_{kn}^* = 4 \sin^2 \left[\frac{\pi}{2} \left(\frac{2k-1}{2n+1} \right) \right] \quad (k = 1, \ldots, n).$$

By using the variance ratio $c = \sigma_x^2/\sigma_v^2 \, (\geq 0)$,[3] we rewrite $-(1/2)L_n$ as

$$L_{1n} = \sum_{k=1}^{n} [\log \sigma_v^2 + \log(a_{kn}^* + c)] + \frac{1}{\sigma_v^2} \sum_{k=1}^{n} \frac{z_k^2}{a_{kn}^* + c}. \qquad (4.9)$$

Since $z_k \sim N(0, a_{kn}^* \sigma_v^2 + \sigma_x^2)$ $(k = 1, \ldots, n)$, the maximum likelihood estimator of σ_v^2 can be represented as

$$\hat{\sigma}_{v.ML}^2 = \frac{1}{n} \sum_{k=1}^{n} \frac{z_k^2}{a_{kn}^* + c} \qquad (4.10)$$

and the concentrated (normalized) log-likelihood function in this simple case is proportional to $-(1/2)$ times

$$L_{1n}(c) = \log[\frac{1}{n} \sum_{k=1}^{n} \frac{z_k^2}{a_{kn}^* + c}] + \frac{1}{n} \sum_{k=1}^{n} \log[a_{kn}^* + c] + 1. \qquad (4.11)$$

Then it may not be straightforward to obtain the maximum likelihood estimator of c because the likelihood function may not be a simple function and the likelihood equation $\frac{\partial L_{1n}(c)}{\partial c} = 0$ is a polynomial function of order $2n - 1$, and as a consequence there are local maximum points with any finite sample data.

As a typical situation we draw the likelihood function with respect to the parameter c and the result of a small simulation in Fig. 4.6 when the true values are $\sigma_v^2 = 0.4$ and $\sigma_x^2 = 0.2$ (Case 4-1). We also show the likelihood function and the result of case when the true values are $\sigma_v^2 = 0.1$ and $\sigma_x^2 = 0.8$ (Case 4-2) as Fig. 4.7. To make a comparison, we give Case 4-3 ($\sigma_v^2 = 0.2, \sigma_x^2 = 0.4$) as Fig. 4.8 and Case 4-4 ($\sigma_v^2 = 0.8, \sigma_x^2 = 0.1$) as Fig. 4.9. The shape of the likelihood function is reasonable in Case 4.1 and Case 4.4 while it is not so in Case 4.2 and Case 4.3. In the latter cases, the likelihood function is rather flat around the maximum point, which means that it does not have much information in a sense.

Case 4-1 and Case 4-4 are the standard ones and we can expect that the ML estimation under the Gaussian assumption gives a reasonable result. However, Case 4-2 and Case 4.3 illustrate some problem on the ML estimation. The likelihood function looks flat over a wide range of the parameter space of c and it causes some computation difficulty to find its maximum. We could interpret the reason why Akaike

---

[3] The present notation c corresponds to $c^{-1/2}$ in Akaike (1989).

Fig. 4.6 Likelihood
function ($n = 100, c = 1/2$)

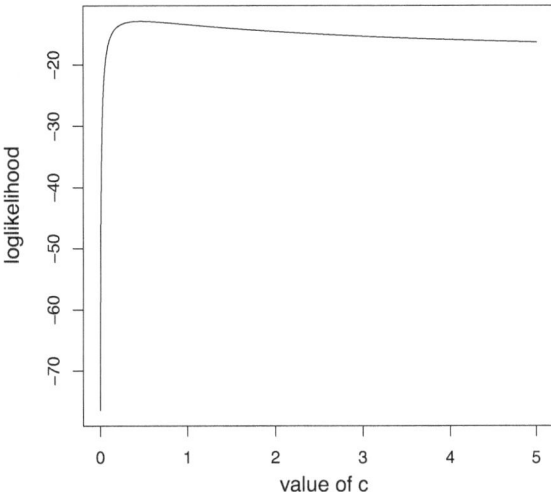

Fig. 4.7 Likelihood
function ($n = 100, c = 8$)

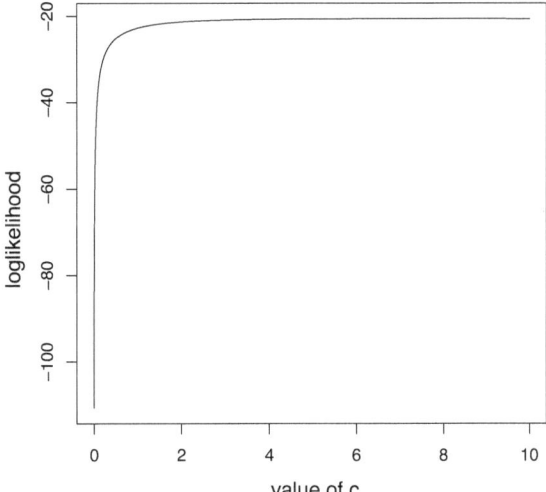

(1989) suggested that we should impose the restriction $0 < c \leq c_u$ for a pre-specified c_u in our setting because of the usefulness of the resulting statistical models of time series filtering. Kitagawa (2021) has developed the DECOMP program,[4] which has a similar restriction on the parameter space. We sometimes obtain the estimated value quite near 1.0 for c when we analyze macroeconomic data such as real (quarterly) GDP in Japan.

---

[4] It is available freely at website: http://jasp.ism.ac.jp/RS-DECOMP (Institute of Statistical Mathematics, Tokyo).

Fig. 4.8 Likelihood
function ($n = 100, c = 2$)

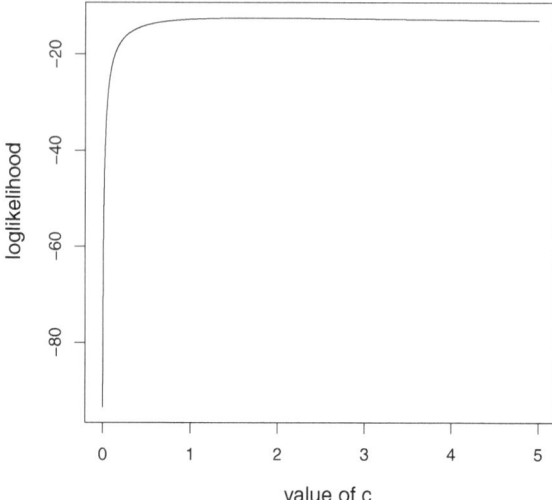

Fig. 4.9 Likelihood
function ($n = 100, c = 1/8$)

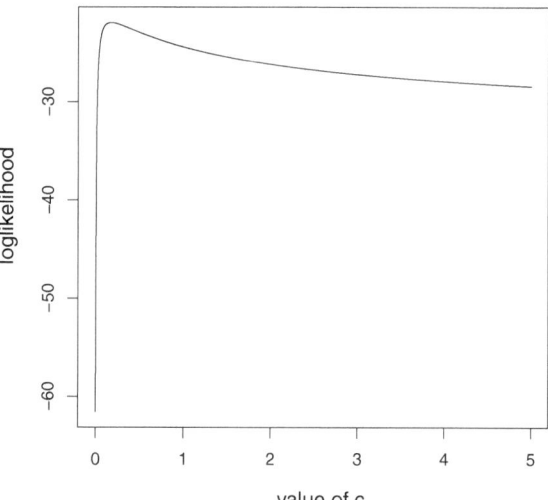

One of interesting aspects of the present problem is the fact the ML method does not necessary give a satisfactory solution meaning a numerically stable solution and it is the case when $p = 1, \sigma_v^2 = 0.1$, and $\sigma_x^2 = 0.8$. Since the likelihood function is near flat over a wide range of the parameter space, it is often difficult to find the maximum in a stable way. In fact, given the finite sample, there is a positive probability of zero when the true value of σ_v^2 is small, but not zero. This can be easily shown because the normalized second-order derivative of the log-likelihood function can be approximated as some negative number when $\tau \ (= 1/c) \to 0$ and the zero point is a local maximum. As an illustration, we give the histogram of the ML

Fig. 4.10 Histogram of ML
($n = 100, c = 8$)

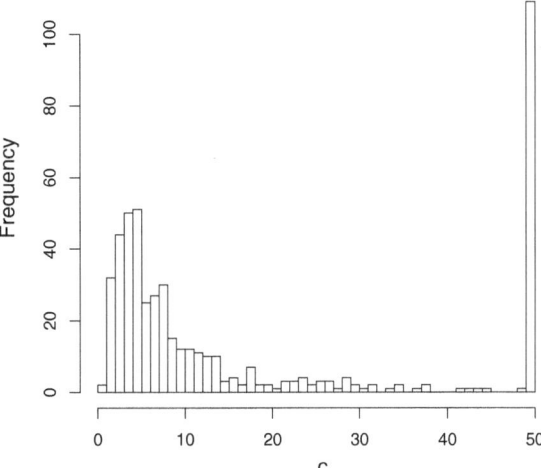

estimation with the restriction $0 < c \le 50$ for the second case as Fig. 4.10 with 500 replications. There is a difficulty to find the global maximum of likelihood function and there is some probability at the boundary point of $c = 50$.

On the other hand, it is certainly possible to approximate the log-likelihood function (4.8) as

$$L_n^{SI} = \sum_{k=1}^{m_n} \log |a_{kn}^* \sigma_v^2 + \sigma_x^2|^{-1/2} - \frac{1}{2} \sum_{k=1}^{m_n} \frac{z_k^2}{a_{kn}^* \sigma_v^2 + \sigma_x^2}. \qquad (4.12)$$

We set the requirement on $m_n/n = o(1)$ because $a_{kn}^* = o(1)$ when $k = 1, \ldots, m_n$ and $m_n, n \to \infty$. Then the SIML estimator σ_x^2 (we call the macro-SIML method) can be given by

$$\hat{\sigma}_{x.SIML}^2 = \frac{1}{m_n} \sum_{k=1}^{m_n} z_k^2 . \qquad (4.13)$$

Since the information on the trend term is separated from the noise term, we expect that the resulting macro-SIML estimation has a robust property.

We note that the macro-SIML estimation of σ_v^2 is not the same as the original (finance) SIML method developed by Kunitomo et al. (2018) because $a_{kn}^* = O(1)$ for $k = n - m_n + 1, \ldots, n$. One way to estimate σ_v^2 is to use the fact that

$$\mathcal{E}\left[\frac{1}{n} \sum_{k=1}^{n} z_k^2 \right] = \sigma_x^2 + \left(\frac{1}{n} \sum_{k=1}^{n} a_{kn}^* \right) \sigma_v^2 \longrightarrow \sigma_x^2 + 2\sigma_v^2 \ (n \to \infty). \qquad (4.14)$$

It is clear that $a^*_{kn} \longrightarrow 0$ when $k_n/n \to 0$ $(n \to \infty)$ and $a^*_{kn} \to 4$ $(k_n/n \to 1, n \to \infty)$. Then a possible SIML estimator σ^2_v can be given by

$$\hat{\sigma}^2_{v.SIML}(1) = \frac{1}{2}\left[\frac{1}{n}\sum_{k=1}^{n} z^2_k - \hat{\sigma}^2_{x.SIML}\right], \tag{4.15}$$

with the restriction of non-negativity.

For the estimation problem of high-frequency financial data, Kunitomo et al. (2018) have suggested to use $\hat{\sigma}^2_{v.SIML} = \frac{1}{l_n}\sum_{k=n-l_n}^{n} a^{-1}_{kn} z^2_k$, where $a_{kn} = na^*_{kn}$ and $l_n = o(n)$ for the high-frequency asymptotics. However, in the present case of Macro-SIML with a^*_{kn} it is straightforward to show that $a^*_{n-l+1,n} \to 4$ as $n \to \infty$ for fixed l, the macro-SIML estimator is given by

$$\hat{\sigma}^2_{v.SIML}(2) = \frac{1}{l_n}\sum_{k=n-l_n+1}^{n} a^{*-1}_{kn} z^2_k - \frac{1}{4}\hat{\sigma}^2_{x.SIML} \tag{4.16}$$

with the restriction of non-negativity.

4.3.2   A Non-stationary Common Trend Case

The difficulty in the ML estimation becomes more serious in the multivariate non-stationary errors-in-variables models. We illustrate the problem of multivariate aspects in a simple, but important formulation. It can be regarded as a simple extension of the so-called reduced rank regression.

Let $\mathbf{y}_i$ be the i-th observation of p-dimensional time series $(i = 1, \ldots, n)$, $\mathbf{y}_i = \mathbf{x}_i + \mathbf{v}_i$, and $\mathbf{Y}_n = (\mathbf{y}'_i)$ be an $n \times p$ $(p > 1)$ matrix of observation. We assume that the vectors $\mathbf{x}_i$ satisfy

$$\mathbf{x}_i = \mathbf{x}_{i-1} + \mathbf{v}_i^{(x)}, \tag{4.17}$$

and $\mathbf{v}_i^{(x)} = \boldsymbol{\pi}\mu^*_i$, $\boldsymbol{\pi}$ is a (non-zero) $p \times 1$ vector, μ^*_i is a sequence of i.i.d. (one-dimensional) random variables[5] following $N(0, \sigma^2_\mu)$, and $\mathbf{v}_i$ are i.i.d. (p-dimensional) random variables following $N_p(\mathbf{0}, \boldsymbol{\Sigma}_v)$ with the variance-covariance (non-singular) matrix $\boldsymbol{\Sigma}_v$. We set $\mathbf{b} = \sigma_\mu \boldsymbol{\pi}$ and $\mathbf{A} = a^*_{kn}\boldsymbol{\Sigma}_v$ and then apply the matrix formulas such that for a positive definite $\mathbf{A}$ and non-zero vector $\mathbf{b}$

$$|\mathbf{A} + \mathbf{bb}'| = |\mathbf{A}|[1 + \mathbf{b}'\mathbf{A}^{-1}\mathbf{b}] \tag{4.18}$$

and

$$[\mathbf{A} + \mathbf{bb}']^{-1} = \mathbf{A}^{-1} - \mathbf{A}^{-1}\mathbf{b}[1 + \mathbf{b}'\mathbf{A}^{-1}\mathbf{b}]^{-1}\mathbf{b}'\mathbf{A}^{-1} \tag{4.19}$$

---

[5] The notation μ^*_i is different from μ_i and $\mu^*_i = \Delta\mu_i$ $(i = 2, \ldots, n)$ in Kunitomo and Sato (2017).

for $\Sigma_x = \mathbf{b}\mathbf{b}'$.

The likelihood function L_n is proportional to $(-1/2)$ times

$$
L_{1n} = \sum_{k=1}^{n} \left[\log |a_{kn}^* \Sigma_v| + \log(1 + a_{kn}^{*-1} \mathbf{b}' \Sigma_v^{-1} \mathbf{b}) \right.
$$
$$
\left. + a_{kn}^{*-1} \mathbf{z}_k' \Sigma_v^{-1} \mathbf{z}_k - \frac{a_{kn}^{*-1} (\mathbf{z}_k' \Sigma_v^{-1} \mathbf{b})^2}{a_{kn}^* + \mathbf{b}' \Sigma_v^{-1} \mathbf{b}} \right]
$$
$$
= \sum_{k=1}^{n} \log |a_{kn}^* \Sigma_v| + \sum_{k=1}^{n} a_{kn}^{*-1} \mathbf{z}_k' \Sigma_v^{-1} \mathbf{z}_k
$$
$$
+ \sum_{k=1}^{n} \left[\log(1 + a_{kn}^{*-1} c) - \frac{a_{kn}^{*-1} (\mathbf{z}_k' \Sigma_v^{-1} \mathbf{b})^2}{a_{kn}^* + c} \right],
$$

where we denote

$$
c = \sigma_\mu^2 \pi' \Sigma_v^{-1} \pi . \tag{4.20}
$$

We need a normalization for vector π and one possibility is to take $\pi' = (1, -\theta_2')$, but there can be other possibility.

Remark 4.1 When $p = 2$, we take $\beta' = (1, -\beta_2)$ and $\pi' \beta = 0$ with a normalization. Then we can interpret $\beta' \mathbf{y}_i = \beta' \mathbf{v}_i \, (= u_i)$ (the rank of π is 1) as the structural equation in time series econometrics. It is because $\mathbf{y}_i$ is an I(1) vector and while $\beta' \mathbf{y}_i = u_i$ is an I(0) variable, where d in I(d) ($d = 0, 1$) is the integration order of time series.

As an intuitive way to simplify the present problem of statistical relationship among non-stationary variables and to obtain the solution is to use the moment condition that

$$
\mathbf{E}[\mathbf{z}_k \mathbf{z}_k'] = \Sigma_x + o(1) \quad \text{for } k = 1, \ldots, m_n
$$

and

$$
\mathbf{E}[a_{kn}^{*-1} \mathbf{z}_k \mathbf{z}_k'] = \Sigma_v + \frac{1}{4} \Sigma_x + o(1) \quad \text{for } k = n + 1 - m_n, \ldots, n.
$$

In the present case, the rank of matrix Σ_x is one while the matrix Σ_v has a full rank. When $p = 2$ in particular, we can find a vector β uniquely such that $\Sigma_x \beta = 0$ with a normalization (see Sect. 4.5 for more general cases).

To estimate the structural equation vector β, then, it may be natural to consider the characteristic equation

$$
\left[\hat{\Sigma}_{x.SIML} - \lambda \hat{\Sigma}_{v.SIML} \right] \hat{\beta}_{SIML} = \mathbf{0}, \tag{4.21}
$$

and

$$\hat{\Sigma}_{x.SIML} = \frac{1}{m_n} \sum_{k=1}^{m_n} \mathbf{z}_k \mathbf{z}_k', \tag{4.22}$$

$$\hat{\Sigma}_{v.SIML}(1) = \frac{1}{2}[\frac{1}{n} \sum_{k=1}^{n} \mathbf{z}_k \mathbf{z}_k' - \hat{\Sigma}_{x.SIML}], \tag{4.23}$$

or

$$\hat{\Sigma}_{v.SIML}(2) = \frac{1}{l_n} \sum_{k=n+1-l_n}^{n} a_{kn}^{*-1} \mathbf{z}_k \mathbf{z}_k' - \frac{1}{4} \hat{\Sigma}_{x.SIML}, \tag{4.24}$$

where $\mathbf{Z}_n = (\mathbf{z}_k') = \mathbf{P}_n \mathbf{C}_n^{-1} (\mathbf{Y}_n - \mathbf{1}_n \bar{\mathbf{y}}_0)$ as (6), λ is the (scalar) eigenvalue, and $\hat{\Sigma}_{x.SIML}$ and $\hat{\Sigma}_{v.SIML}$ are the SIML estimators of Σ_x and Σ_v, respectively. We require the condition that these estimators of variance-covariance matrices are non-negative definite.

Because the rank of Σ_x is degenerated and it is one in the present case, it may be natural to use the smaller eigenvalue, say, λ_1. Then the resulting characteristic vector $\hat{\boldsymbol{\beta}}_{SIML}$ is called the SIML estimator of $\boldsymbol{\beta}$ because of (4.22). Since the estimated variance-covariance matrix of Σ_v should be positive definite, we may have instability in some cases if we use (4.23) or (4.24) without any restriction as the non-negative definiteness.

A simplified (consistent) estimation may be given by

$$\hat{\Sigma}_{x.SIML} \times \hat{\boldsymbol{\beta}}_{SIL} = \mathbf{0}, \tag{4.25}$$

that is,

$$\hat{\Sigma}_{x.SIML} \times \begin{bmatrix} 1 \\ -\hat{\boldsymbol{\beta}}_{2.SIL} \end{bmatrix} = \mathbf{0}. \tag{4.26}$$

We can solve as

$$\hat{\boldsymbol{\beta}}_{2.SIL} = \hat{\Sigma}_{22x.SIML}^{-1} \hat{\Sigma}_{21x.SIML}, \tag{4.27}$$

where $\hat{\Sigma}_{22x.SIML}$ and $\hat{\Sigma}_{21x.SIML}$ are the (2,2)-element and (2,1)-element of $\hat{\Sigma}_{x.SIML}$, respectively. ($\hat{\Sigma}_{22x.SIML}$ is positive with probability one.)

It is the least squares method for the transformed variables $\mathbf{z}_k$ ($k = 1, \ldots, m$) and hence we call the separating information least squares (SILS) estimator.

Remark 4.2 We note that there were extensive discussions on alternative estimation methods including the similar form as the SIML and the SISL estimators for the errors-in-variables models and the single structural equation econometric models for independent observations. Some improvements on the finite sample properties may be possible. See Anderson (1984), Anderson and Takemura (1986), and Fuller (1987).

Fig. 4.11 Likelihood
function of θ $(n = 1,000)$

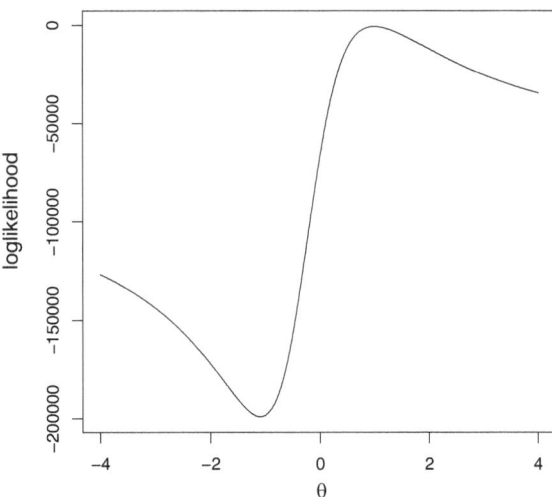

4.4 Gaussian Likelihood Function and Estimation Methods

It may be natural to apply the general parametric principle of the maximum likelihood
estimation method (ML). One of interesting aspects of the present problem is the
fact that the ML method does not necessarily give a satisfactory solution.

As a two-dimension example, we use the example in Sect. 4.3.2 and set the true
parameter values as $\sigma_\mu^2 = 0.4, \theta = 1.0$, and

$$\Sigma_v = \begin{pmatrix} 0.45 & 0.23 \\ 0.23 & 0.4 \end{pmatrix}, \quad \Sigma_x = \sigma_\mu^2 \pi \pi', \quad \pi = \begin{pmatrix} 1 \\ -\theta \end{pmatrix}.$$

Then we generate a set of simulated observations and we have drawn the Gaussian
likelihood functions of θ in Figs. 4.11 and 4.12 when $1,000$, given the true values
for other parameters. It is possible to attain the maximum point of the likelihood
function locally as shown in Fig. 4.11. It suggests that there is a global maximization
problem because we need the right starting point for the maximization. We have
investigated the likelihood function in different cases and found that the likelihood
function could have some non-concave forms in some cases as illustrated by Fig. 4.12.
We found that the Gaussian likelihood function is nearly flat with respect to the
correlation coefficient parameter and the maximization may be difficult with respect
to correlation of the noise terms. These are some of important consequences in the
non-stationary errors-in-variables models.

It is interesting to see what happens if the Gaussian assumption is not true and as an
illustration we have drawn one wrong likelihood function in Fig. 4.13 on this problem.
We generated the random variables followed by the uniform distribution on $[-2, +2]$
(i.e., $\mathbf{v}_i^{(x)} = (v_{ji}^{(x)})$), the distribution of $\mathbf{v}_i$ is normal and the correlation coefficient ρ

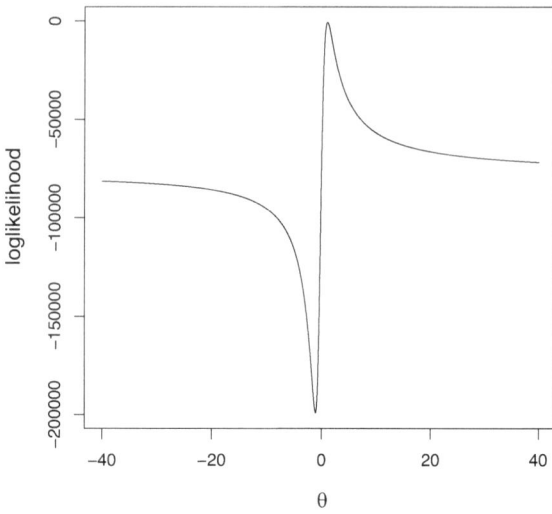

Fig. 4.12 Likelihood
function of θ ($n = 1,000$)

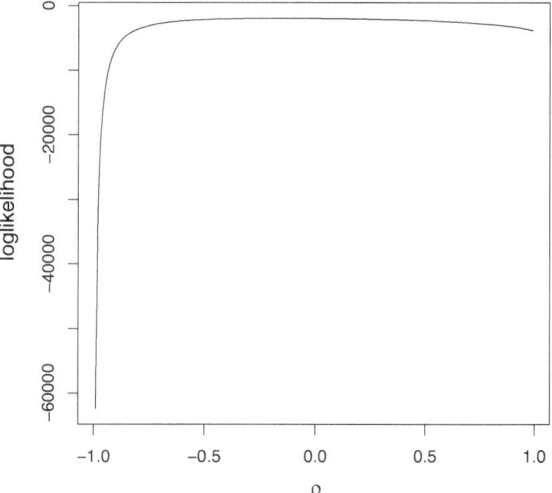

Fig. 4.13 Wrong likelihood
function of ρ ($n = 1,000$)

is 0.3. As Fig. 4.13 (ρ, which is the correlation coefficient among measurement
errors $\mathbf{v}_i$) suggests, the ML estimation of the variance-covariance matrix of trend
components crucially depends on the assumption of Gaussianity as we had expected.
Hence we have a risk to use the ML computation to investigate the relationships
among hidden trend variables unless we have the strong support for the Gaussianity
of data.

Now we investigate the asymptotic properties of $(-1)\times$ the log-likelihood function and the estimation methods (the ML and SIML estimators) when $\boldsymbol{\Sigma}_x = \mathbf{bb}'$ ($\mathbf{b} \neq \mathbf{0}$) (i.e., the rank of $\boldsymbol{\Sigma}_x$ is 1) and $p \geq 2$. We normalize the Gaussian log-likelihood function by $-(1/n)$ and rewrite

$$L_{1n}^* = \frac{1}{n} \sum_{k=1}^{n} |a_{kn}^* \boldsymbol{\Sigma}_v|$$

$$+ \frac{p}{n} \sum_{k=1}^{n} \log(1 + a_{kn}^{*-1} c)$$

$$+ \frac{1}{n} \sum_{k=1}^{n} a_{kn}^{*-1} \mathbf{tr} \left[\boldsymbol{\Sigma}_v^{-1} \left(\boldsymbol{\Sigma}_v - \frac{1}{a_{kn}^* + c} \mathbf{bb}' \right) \right.$$

$$\left. \boldsymbol{\Sigma}_v^{-1} (\mathbf{z}_k \mathbf{z}_k' - (a_{kn}^* \boldsymbol{\Sigma}_v(\theta_0) + \mathbf{b}(\theta_0)\mathbf{b}(\theta_0)')) \right]$$

$$+ \frac{1}{n} \sum_{k=1}^{n} a_{kn}^{*-1} \mathbf{tr} \left[\boldsymbol{\Sigma}_v^{-1} \left(\boldsymbol{\Sigma}_v - \frac{1}{a_{kn}^* + c} \mathbf{bb}' \right) \boldsymbol{\Sigma}_v^{-1} (a_{kn}^* \boldsymbol{\Sigma}_v(\theta_0) + \mathbf{b}(\theta_0)\mathbf{b}(\theta_0)') \right]$$

$$= \frac{1}{n} \sum_{k=1}^{n} |a_{kn}^* \boldsymbol{\Sigma}_v| + \frac{p}{n} \sum_{k=1}^{n} \log(1 + a_{kn}^{*-1} c) + L_{12n}^* + L_{13n}^* \quad (say), \qquad (4.28)$$

where $\boldsymbol{\Sigma}_v(\theta_0)$ and $\mathbf{b}(\theta_0)$ are $\boldsymbol{\Sigma}_v$ and $\mathbf{b}$, respectively, evaluated at the true parameter values.

We prepare the next lemma.

Lemma 4.1 *Let a $p \times p$ random vector $\mathbf{z}_k$ follows $N_p(\mathbf{0}, \mathbf{Q})$. Then for any $p \times p$ matrices $\mathbf{A}_k$,*

$$\mathbf{E}[(\mathbf{tr}(\mathbf{A}_k \mathbf{z}_k \mathbf{z}_k'))^2] = [\mathbf{tr}(\mathbf{A}_k \mathbf{Q})]^2 + 2\mathbf{tr}(\mathbf{A}_k \mathbf{Q} \mathbf{A}_k \mathbf{Q}). \qquad (4.29)$$

By using this lemma, it is straightforward to see that as $n \longrightarrow \infty$ the third term of (4.28) converges to

$$L_{12n}^* \overset{p}{\longrightarrow} 0. \qquad (4.30)$$

Then we can establish the next result on the ML estimation by evaluating the remaining terms of L_{1n}^* as in the Appendix of this chapter. Although we expect that the ML estimation under the Gaussian assumption has good asymptotic properties, we could not find any proof in the present setting and we have given it in the Appendix of this chapter.

Theorem 4.1 *For $p \geq 2$ and the rank of $\boldsymbol{\Sigma}_x$ being 1, we set $\boldsymbol{\Sigma}_x = \mathbf{bb}'$ in (4.1). Assume that $\mathbf{v}_i$ $(i = 1, \ldots, n)$ are the sequence of i.i.d. random vectors and $\boldsymbol{\Sigma}_v$ is a positive definite matrix. Then under the assumption of Gaussian distributions of $\mathbf{v}_i$ and $\mathbf{w}_i^{(x)}$, the maximum likelihood estimator of $\mathbf{b}$ and $\boldsymbol{\Sigma}_v$ are consistent as $n \longrightarrow \infty$.*

Remark 4.3 It should be noted that in time series econometrics it has been known that coefficient parameter vector $\boldsymbol{\beta}$ can be estimated by using the standard regression method if the observed variables are co-integrated. Johansen (1995) had developed the ML method without any noise term and investigated the estimation and inference of the co-integrating vectors.

We should mention the fact that the SIML estimator does have not only consistency as well as the asymptotic normality under the standard regularity conditions such as the fourth-order moments without the Gaussian assumption. See Sect. 5.4 and the Appendix of Chap. 5.

There has not been any finite sample result on the estimation methods for the non-stationary time series models with errors-in-variables. Since there are many situations with macroeconomic variables that we observe the non-stationarity with measurement errors, it is worthwhile to investigate the related issues by using simulations. As we have seen in Sect. 4.3 the ML estimator has finite sample instability and we only report the finite sample properties of the SIML estimation in this section.

We have set $\sigma_\mu^2 = 1$, $\sigma_v^2 = 0.5, 2$, or 4, $\beta_2 = 1.5$, where we summarize our setting of simulations as

$$
\Sigma_x = \begin{pmatrix} \Sigma_{x,11} & \Sigma_{x,12} \\ \Sigma_{x,12} & \Sigma_{x,22} \end{pmatrix} = \sigma_\mu^2 \begin{pmatrix} \beta_2 \\ 1 \end{pmatrix} (\beta_2, 1),
$$

$$
\Sigma_v = \begin{pmatrix} \Sigma_{v,11} & \Sigma_{v,12} \\ \Sigma_{v,12} & \Sigma_{v,22} \end{pmatrix}.
$$

(The parametrization is slightly different from Figs. 4.11 and 4.12.) We took the cases when $n = 80$ or 400 and $m_n = [n^\alpha]$ with $\alpha = 0.6$ or 0.7, and the number of Monte Carlo repetitions is 1,500 in each case. From our simulations we summarize mail results as Table 4.1. In Table 4.1 the numbers inside the parentheses are the standard deviation of estimators calculated by our simulations. We found that the SIML estimator of trend variance-covariance estimates $\Sigma_{x.ij}$ $(i, j = 1, 2)$ have reasonable finite sample properties. We also give the SIML estimator of noise variance-covariance estimates $\Sigma_{v.ij}$ $(i, j = 1, 2)$, which is based on (4.24) are reasonable. The SILS estimate of β_2 is biased a little in comparison with the SIML estimates, but the former has smaller sample variance than the latter.

We have done many simulations, but the results are similar with Table 4.1 in the present formulation. There are several general findings, which are summarized as follows. First, on the effects of sample sizes for the performance of the estimators of the SIML estimation, it becomes better as the sample size increases as we had expected. Second, when the variances of noises are small, both the SIL estimator (4.27) and the SIML estimator (4.22) give reasonable estimates on the coefficient parameter, the former is slightly biased toward zero while the latter has some correction of bias. The variability of the SIML estimate in terms of simulation variance is slightly larger than that of the SIL estimate. Third, when the variances of noises are not small, the SILS estimator has a significant bias.

Table 4.1 Finite sample properties of SIML

σ_v^2	α	n	$\Sigma_{x,11}$	$\Sigma_{x,12}$	$\Sigma_{x,22}$	$\Sigma_{v,11}(1)$	$\Sigma_{v,12}(1)$	$\Sigma_{v,22}(1)$	$\beta_{2.SIL}$	$\beta_{2.SIML}$
0.5	0.6	80	2.263	1.475	1.038	0.495	0.010	0.499	1.420	1.528
			(0.904)	(0.602)	(0.415)	(0.445)	(0.305)	(0.318)	(0.114)	(0.178)
		400	2.239	1.484	1.006	0.512	0.009	0.501	1.475	1.502
			(0.542)	(0.362)	(0.244)	(0.283)	(0.185)	(0.186)	(0.038)	(0.038)
	0.7	80	2.294	1.454	1.086	0.521	0.030	0.496	1.339	1.554
			(0.706)	(0.469)	(0.337)	(0.386)	(0.250)	(0.251)	(0.131)	(0.269)
		400	2.296	1.499	1.044	0.498	0.007	0.494	1.436	1.502
			(0.393)	(0.263)	(0.182)	(0.214)	(0.137)	(0.136)	(0.047)	(0.054)
2	0.6	80	2.378	1.438	1.163	1.992	0.006	1.922	1.233	1.630
			(0.948)	(0.617)	(0.455)	(1.020)	(0.699)	(0.852)	(0.231)	(0.810)
		400	2.318	1.500	1.060	1.981	0.006	1.991	1.415	1.504
			(0.534)	(0.352)	(0.245)	(0.615)	(0.411)	(0.535)	(0.077)	(0.082)
	0.7	80	2.629	1.452	1.438	1.943	0.032	1.944	1.017	1.701
			(0.833)	(0.543)	(0.471)	(0.816)	(0.570)	(0.709)	(0.230)	(0.901)
		400	2.410	1.479	1.166	1.975	0.007	1.946	1.267	1.514
			(0.534)	(0.440)	(0.287)	(0.465)	(0.309)	(0.407)	(0.093)	(0.123)
4	0.6	80	2.639	1.469	1.384	3.927	-0.010	3.990	1.072	1.702
			(1.040)	(0.669)	(0.547)	(1.837)	(1.214)	(1.670)	(0.296)	(1.042)
		400	2.377	1.503	1.127	3.933	-0.008	3.965	1.334	1.514
			(0.558)	(0.368)	(0.267)	(1.099)	(0.727)	(1.025)	(0.105)	(0.132)
	0.7	80	3.118	1.457	1.885	3.777	0.065	3.874	0.787	1.846
			(1.005)	(0.636)	(0.630)	(1.427)	(0.963)	(1.311)	(0.274)	(1.452)
		400	2.601	1.483	1.357	3.937	0.010	3.908	1.095	1.519
			(0.451)	(0.298)	(0.249)	(0.806)	(0.550)	(0.727)	(0.119)	(0.197)

To summarize our simulations, the finite sample performance of the SIML estimation gives reasonable performances as the asymptotic theory has suggested as in the previous sections.

4.5 Extensions

There can be extensions of the problem we have been investigating. We discuss two of them rather briefly.

4.5.1 Reduced Rank Condition

For the multivariate non-stationary (economic) time series, there are possibilities of co-integration in trends. In our framework, it may be interesting to consider the general case of reduced rank cases when

$$\text{rank}[\mathbf{\Sigma}_x] = q_x \ , \quad 1 \le q_x < p \ , \tag{4.31}$$

where we can represent $\mathbf{\Sigma}_x = \mathbf{B}\mathbf{B}'$ and $\mathbf{B}$ is a $p \times q_x$ matrix (its rank is q_x).

Then the example in Sect. 4.3.2 corresponds to the case when $q_x = 1 < p$ and $p \ge 2$.

In the more general cases, however, there is a parametrization problem for the $p \times p$ matrix $\mathbf{\Sigma}_x$, whose rank is q_x ($1 \le q_x < p, \ p \ge 2$). Then we can take a $p \times r_x$ ($r_x = p - q_x$) matrix $\boldsymbol{\beta}$ such that $\mathbf{B}'\boldsymbol{\beta} = \mathbf{O}$ and a normalization such as $\boldsymbol{\beta}'\mathbf{\Sigma}_v^{-1}\boldsymbol{\beta} =$ (diag c_{ii}) for instance. The algebra of Sect. 4.3.2 can be extended by using the matrix formulae such that for a positive definite matrix $\mathbf{A}$, a $p \times q_x$ matrix $\mathbf{B}$ (the rank is q_x), and a $p \times r_x$ matrix $\boldsymbol{\beta}$ (the rank is r_x), we have $\mathbf{\Sigma}_x = \mathbf{B}\mathbf{B}'$, $|\mathbf{A} + \mathbf{B}\mathbf{B}'| = |\mathbf{A}||\mathbf{I}_{q_x} + \mathbf{B}'\mathbf{A}^{-1}\mathbf{B}|$, and

$$[\mathbf{A} + \mathbf{B}\mathbf{B}']^{-1} = \mathbf{A}^{-1} - \mathbf{A}^{-1}\mathbf{B}[\mathbf{I}_{q_x} + \mathbf{B}'\mathbf{A}^{-1}\mathbf{B}]^{-1}\mathbf{B}'\mathbf{A}^{-1}. \tag{4.32}$$

Hence we can use the characteristic vectors with the r_x smaller characteristic roots of the equation.

The special case when $q_x = p - 1$ has attracted special attention among econometricians or economists because there exists a unique vector $\boldsymbol{\beta}$ such that $\mathbf{\Sigma}_x\boldsymbol{\beta} = \mathbf{0}$ with a normalization. It has been called the single structural equation in traditional econometrics or the co-integrated relation in time series econometrics and we can extend the developments of Sect. 4.4. When $p > 2$ and $0 < q_x < p - 1$, there are r_x ($= p - q_x$) co-integrated vectors satisfying $\mathbf{B}'\boldsymbol{\beta} = \mathbf{O}$ and then there is an identification problem on the vectors consisting of $\boldsymbol{\beta}$.

Remark 4.4 The problem of reduced rank condition was a central issue in Anderson (1984) for the case of independent observations in statistical multivariate analysis. For the non-stationary time series without measurement errors, Engle and Granger (1987) and Johansen (1995) have developed the statistical inference and they called the resulting equations from the non-stationary multiple time series as co-integrated relations. It is possible to develop the testing procedure of the rank condition for hidden non-stationary components.

4.5.2 Higher Order Integrated Processes

In some cases the second-order (or higher order) differencing may be often appropriate for modeling economic time series. Although the likelihood function can be complicated in general, we can develop the SIML estimation when $p \ge 1$ and $d = 2$, where

$$\Delta^d \mathbf{x}_i = \mathbf{v}_i^{(x)} \ , \tag{4.33}$$

$\mathbf{E}[\mathbf{v}_i^{(x)}] = \mathbf{0}$, and $\mathbf{E}[\mathbf{v}_i^{(x)}\mathbf{v}_i^{(x)'}] = \mathbf{\Sigma}_x$.

We use the $\mathbf{K}_n$−transformation that from the observation matrix $\mathbf{Y}_n$ to $\mathbf{Z}_n^{(2)}$ ($= (\mathbf{z}_k^{(2)'})$) by

$$\mathbf{Z}_n^{(2)} = (\mathbf{z}_k^{(2)'}) = \mathbf{K}_n \left(\mathbf{Y}_n - \bar{\mathbf{Y}}_0 \right), \mathbf{K}_n = \mathbf{P}_n \mathbf{C}_n^{-2}. \tag{4.34}$$

Then the separating information maximum likelihood (SIML) estimator of $\hat{\boldsymbol{\Sigma}}_x$ in this case can be defined by

$$\hat{\boldsymbol{\Sigma}}_{x,SIML} = \frac{1}{m_n} \sum_{k=1}^{m_n} \mathbf{z}_k^{(2)} \mathbf{z}_k^{(2)'}. \tag{4.35}$$

We prepare the next Lemma.

Lemma 4.2 *Let*

$$\mathbf{K}_n = (b_{ij}^{(2)}) = \mathbf{P}_n \mathbf{C}_n^{-2}. \tag{4.36}$$

Then for $i, i' = 1, \ldots, m_n$, the dominant term of $\sum_{j=1}^n b_{ij}^{(2)} b_{i',j}^{(2)}$ is

$$\delta(i, i') \left[2 \sin \left(\frac{\pi}{2} \frac{2i-1}{2n+1} \right) \right]^4. \tag{4.37}$$

By using the above lemma, it is straightforward to obtain the next result for the case of $d \geq 1$. We omit the proof because it is essentially the same to the case when $d = 1$ except Lemma 4.36. It is clear that the result holds for a positive integer d.

Theorem 4.2 *Assume $p \geq 1$, $d = 2$, and $m_n/n \longrightarrow 0$ as $n \longrightarrow \infty$, where $m_n = [n^\alpha]$ $(0 < \alpha < 1)$. Under the assumption of existence of fourth-order moments, the SIML estimator of $\boldsymbol{\Sigma}_x$ is consistent with $0 < \alpha < 1$ and asymptotically normal with $0 < \alpha < 0.8$ as $n \longrightarrow \infty$.*

It should be important to note that the diagonal elements a_{kn}^* $(k = 1, \ldots, n)$ should be modified to

$$a_{kn}^{(2)} = [2 \sin \frac{\pi}{2n+1} (k - \frac{1}{2})]^4 \tag{4.38}$$

in the present case and we need the corresponding bias correction for estimating the variance-covariance matrix $\boldsymbol{\Sigma}_v$.

Remark 4.5 Akaike (1989) and Kitagawa (2021) have proposed to use the ML estimation with some restriction for the filtering problem of the similar models when $d \geq 1$. The SIML estimation would be a useful tool for the state space modeling of non-stationary multivariate time series because it does not have any computational problem without any restriction and it has an asymptotic robustness. The SIML method may give reasonable estimates not only for the coefficient parameters, but also the variance-covariance matrices even if we do not know $d = 1, 2$ by taking m_n appropriately.

4.6   Concluding Remarks

In this chapter, we have examined alternative estimation methods for the non-stationary errors-in-variables models. We first illustrated the reason why the presence of noise term in the non-stationary time series, even if it was small, forces to change the standard way of thinking in time series analysis. Subsequently, we discussed the finite sample and large sample properties of the ML and SIML estimation methods for the non-stationary errors-in-variables models when there are non-stationary trends and noise components (or the measurement errors). We have found that the Gaussian likelihood function has a non-concave shape in certain instances and the ML method works when the Gaussianity of the non-stationary and stationary components holds with some restriction such as the signal-noise variance ratio in the parameter space. The SIML estimation has the asymptotic robust properties under general conditions of existence of fourth-order moments for consistency and asymptotic normality. We have investigated the conditions for the asymptotic properties of the ML and SIML estimators as well as some simulations. The SIML method yields reasonable estimates for the coefficient parameters when the random variables do not necessarily follow the Gaussian distribution and we do not have much knowledge on the value of the variance ratio c by taking m_n appropriately. Since we usually do not have information on the magnitude of the signal-noise variance ratio and the precise distribution of noises in advance when we observe non-stationary data, it is important to use the statistical method which is not contingent on these conditions for practical purpose.

Further extensions and some discussions are presented in Sects. 4.3 and 4.5. It would be interesting to ascertain whether the results presented in this chapter remain valid when non-stationary seasonal components are present. There can be many applications of the methods we have discussed because many economists are interested in macroeconomic data and their relations by using seasonally adjusted data. Another issue may be that some econometricians have relied on the large sample asymptotic theory and used the approximations based on Brownian motions, which are the limits of random walks. Although there are some procedures based on Quadratic Variation (see (4.4)) of the observed data to estimate the variance-covariances of innovations and the long-run variance-covariances, there is a fundamental problem as illustrated by examples in Sect. 4.3.

If we ignore the presence of noise components and/or measurement components, there could be serious problems. For example, the actual sample size of macroeconomic data is usually not large and often we have measurement errors. In such situations we should be careful to use such asymptotic theory and we need to investigate the finite sample properties of alternative estimation techniques and their improvements.

Appendix of this Chapter: Mathematical Derivations

In this Appendix, we give some details of the derivations omitted in the previous sections.

Proof of Lemma 4.1: When we have $\mathbf{z}_k = (z_{ik}) \sim N_p(\mathbf{0}, \mathbf{Q})$, we can use the formula of the fourth-order moments as

$$\mathbf{E}[z_{ik} z_{jk} z_{i'k} z_{j'k}] = q_{ij} q_{i'j'} + q_{ii'} q_{jj'} + q_{ij'} q_{i'j} , \tag{4.39}$$

where $\mathbf{Q} = (q_{ij})$. Then, it is straightforward to obtain the results. (Q.E.D.)

Proof of Theorem 4.1: By using Lemma 4.1, it is possible to obtain the variance of L_{12n}^*, which converges to 0 as $n \longrightarrow \infty$. We set

$$\mathbf{A}_k = a_{kn}^{*-1} \mathbf{\Sigma}_v^{-1} \left[\mathbf{\Sigma}_v - \frac{1}{a_{kn}^* + c} \mathbf{b} \mathbf{b}' \right] \mathbf{\Sigma}_v^{-1}$$

and

$$\mathbf{Q}_0 = a_{kn}^* \mathbf{\Sigma}_v(\theta_0) + \mathbf{b}(\theta_0) \mathbf{b}(\theta_0)' .$$

Then

$$\mathrm{tr}[\mathbf{A}_k \mathbf{Q}_0] = \mathrm{tr} \left(a_{kn}^{*-1} \mathbf{\Sigma}_v^{-1} \left[\mathbf{\Sigma}_v - \frac{1}{a_{kn}^* + c} \mathbf{b} \mathbf{b}' \right] \mathbf{\Sigma}_v^{-1} [a_{kn}^* \mathbf{\Sigma}_v(\theta_0) + \mathbf{b}(\theta_0) \mathbf{b}(\theta_0)'] \right) .$$

If $\mathbf{Q} = \mathbf{Q}_0$, then we have

$$\mathrm{tr}(\mathbf{A}_k \mathbf{Q}_0) = \mathrm{tr}(\mathbf{I}_p) = p .$$

Also we find

$$\mathbf{E} \left[\sum_{k=1}^n \mathrm{tr} \mathbf{A}_k (\mathbf{z}_k \mathbf{z}_k' - \mathbf{Q}_0) \right]^2 = \mathbf{E} \left[\sum_{k,k'=1}^n \mathrm{tr}[\mathbf{A}_k (\mathbf{z}_k \mathbf{z}_k' - \mathbf{Q}_0)] \mathrm{tr}[\mathbf{A}_{k'} (\mathbf{z}_{k'} \mathbf{z}_{k'}' - \mathbf{Q}_0)] \right]$$

$$= \left[\sum_{k=1}^n \mathcal{E}[(\mathbf{z}_k' \mathbf{A}_k \mathbf{z}_k)^2 - (\mathrm{tr} \mathbf{A}_k \mathbf{Q}_0)^2] \right]$$

$$= \sum_{k=1}^n 2\mathrm{tr}(\mathbf{A}_k \mathbf{Q}_0 \mathbf{A}_k \mathbf{Q}_0). \tag{4.40}$$

If $\mathbf{Q} = \mathbf{Q}_0$, then it is $2pn$.

Next, we investigate the first term and last term of L_{1n}^* in detail. After simple algebra, L_{13n}^* can be re-expressed as

$$L_{13n}^* = \mathrm{tr}\,\boldsymbol{\Sigma}_v^{-1}\boldsymbol{\Sigma}(\theta_0) + \frac{1}{n}\sum_{k=1}^{n} a_{kn}^{*-1}\mathbf{b}(\theta_0)'\boldsymbol{\Sigma}_v^{-1}\mathbf{b}(\theta_0)$$

$$-\frac{1}{n}\sum_{k=1}^{n}\frac{1}{a_{kn}^*+c}\mathbf{b}'[\boldsymbol{\Sigma}_v^{-1}\boldsymbol{\Sigma}(\theta)\boldsymbol{\Sigma}_v^{-1} + a_{kn}^{*-1}\boldsymbol{\Sigma}_v^{-1}\mathbf{b}(\theta_0)\mathbf{b}(\theta_0)'\boldsymbol{\Sigma}_v^{-1}]\mathbf{b}$$

$$= \mathrm{tr}\,\boldsymbol{\Sigma}_v^{-1}\boldsymbol{\Sigma}(\theta_0)$$

$$+\frac{1}{n}\sum_{k=1}^{n}\frac{1}{a_{kn}^*(a_{kn}^*+c)}\left[(a_{kn}^*+c)\mathbf{b}(\theta_0)'\boldsymbol{\Sigma}_v^{-1}\mathbf{b}(\theta_0)\right.$$

$$\left.-a_{kn}^*\mathbf{b}'\boldsymbol{\Sigma}_v^{-1}\boldsymbol{\Sigma}_v(\theta_0)\boldsymbol{\Sigma}_v^{-1}\mathbf{b} - (\mathbf{b}'\boldsymbol{\Sigma}_v^{-1}\mathbf{b}(\theta))^2\right]$$

$$= \mathrm{tr}\,\boldsymbol{\Sigma}_v^{-1}\boldsymbol{\Sigma}(\theta_0)$$

$$+\frac{1}{n}\sum_{k=1}^{n}\frac{1}{a_{kn}^*(a_{kn}^*+c)}\left[a_{kn}^*\mathbf{b}(\theta_0)'\boldsymbol{\Sigma}_v^{-1}(\boldsymbol{\Sigma}_v - \boldsymbol{\Sigma}_v(\theta_0))\boldsymbol{\Sigma}_v^{-1}\mathbf{b}(\theta_0)\right.$$

$$\left.+\mathbf{b}(\theta_0)'\boldsymbol{\Sigma}_v^{-1}\mathbf{b}(\theta_0)\mathbf{b}'\boldsymbol{\Sigma}_v^{-1}\mathbf{b} - (\mathbf{b}'\boldsymbol{\Sigma}_v^{-1}\mathbf{b}(\theta_0))^2\right]$$

because $c = \mathbf{b}'\boldsymbol{\Sigma}_v^{-1}\mathbf{b}$.

The last term is non-negative because of Cauchy-Schwarz's inequality and its minimum occurs at $\mathbf{b} = \mathbf{b}(\theta_0)$ because $\boldsymbol{\Sigma}_v$ is positive definite. Then we need to evaluate the sum of each term in L_{1n}^*, which are of the same order, as

$$\frac{1}{n}\sum_{k=1}^{n}\log|a_{kn}^*\boldsymbol{\Sigma}_v| + \frac{p}{n}\sum_{k=1}^{n}\log[1 + a_{kn}^{*-1}\mathbf{b}'\boldsymbol{\Sigma}_v^{-1}\mathbf{b}] \tag{4.41}$$

and

$$\mathrm{tr}\,\boldsymbol{\Sigma}_v^{-1}\boldsymbol{\Sigma}_v(\theta_0) + \frac{1}{n}\sum_{k=1}^{n}\frac{\mathbf{b}'\boldsymbol{\Sigma}_v^{-1}(\boldsymbol{\Sigma}_v - \boldsymbol{\Sigma}_v(\theta_0))\boldsymbol{\Sigma}_v^{-1}\mathbf{b}}{a_{kn}^* + \mathbf{b}'\boldsymbol{\Sigma}_v^{-1}\mathbf{b}}. \tag{4.42}$$

By using the inversion formula and its determinant in (4.18) and (4.19), the sum of the above two terms can be written as

$$\frac{1}{n}\sum_{k=1}^{n}\log|a_{kn}^*\boldsymbol{\Sigma}_v + \mathbf{b}\mathbf{b}'| + \frac{1}{n}\sum_{k=1}^{n}\mathrm{tr}[a_{kn}^*\boldsymbol{\Sigma}_v + \mathbf{b}'\mathbf{b}]^{-1}[a_{kn}^*\boldsymbol{\Sigma}_v(\theta_0) + \mathbf{b}\mathbf{b}'].$$

By using Lemma 3.2.2 of Anderson (2003), the minimum of each terms occurs at $a_{kn}^*\boldsymbol{\Sigma}_v + \mathbf{b}\mathbf{b}' = a_{kn}^*\boldsymbol{\Sigma}_v(\theta_0) + \mathbf{b}\mathbf{b}'$, that is, $\boldsymbol{\Sigma}_v = \boldsymbol{\Sigma}_v(\theta_0)$ under the assumption of positive definiteness. Hence the global minimum of the (minus) log-likelihood function occurs iff $\mathbf{b} = \mathbf{b}(\theta_0)$ and $\boldsymbol{\Sigma}_v = \boldsymbol{\Sigma}_v(\theta_0)$.

The rest of our arguments for the consistency follow from the general arguments for the extremum estimation (see Theorem 4.1.1 of Amemiya (1985), for instance) and we have the desired result. (**Q.E.D.**)

Proof of Lemma 4.2: We set

$$\theta_{kj} = \frac{2\pi}{2n+1}\left(j - \frac{1}{2}\right)\left(k - \frac{1}{2}\right), \quad \theta_k = \frac{2\pi}{2n+1}\left(k - \frac{1}{2}\right), \tag{4.43}$$

and

$$b_{kj}^{(2)} = \frac{1}{\sqrt{2n+1}}[(1 - e^{i\theta_k})^2 e^{i\theta_{kj}} + (1 - e^{-i\theta_k})^2 e^{-i\theta_{kj}}] \tag{4.44}$$

for $j, k = 1, \ldots, n$.

Then by using direct evaluations

$$
\begin{aligned}
(2n+1)\sum_{j=1}^{n}[b_{kj}^{(2)}]^2 &= (1 - e^{i\theta_k})^4 \frac{1 + e^{i\theta_k}}{1 - e^{i2\theta_k}} + (1 - e^{-i\theta_k})^4 \frac{1 + e^{-i\theta_k}}{1 - e^{-i2\theta_k}} \\
&\quad + 2n(1 - e^{i\theta_k})^2(1 - e^{-i\theta_k})^2 \\
&= (1 - e^{i\theta_k})^3 + (1 - e^{-i\theta_k})^3 + 2n(e^{i\frac{\theta_k}{2}} - e^{-i\frac{\theta_k}{2}})^4 \\
&\sim 2n\left[4\sin^2\frac{\theta_k}{2}\right]^2,
\end{aligned}
$$

which is the dominant term.

Also for $k \neq k'$ we find that

$$(2n+1)\sum_{j=1}^{n}[b_{kj}^{(2)}b_{k'j}^{(2)}]$$

$$= (1 - e^{i\theta_k})^2(1 - e^{-i\theta_k})^2 \sum_{j=1}^{n} e^{i(\theta_{kj} + \theta_{k'j})}$$

$$+ (1 - e^{i\theta_k})^2(1 - e^{-i\theta_{k'}})^2 \sum_{j=1}^{n} e^{i(\theta_{kj} + \theta_{k'j})}$$

$$+ (1 - e^{-i\theta_k})^2(1 - e^{i\theta_{k'}})^2 \sum_{j=1}^{n} e^{i(\theta_{kj} - \theta_{k'j})}$$

$$+ (1 - e^{-i\theta_k})^2(1 - e^{-i\theta_{k'}})^2 \sum_{j=1}^{n} e^{-i(\theta_{kj} + \theta_{k'j})}$$

$$= (I) + (II) + (III) + (IV) \quad (say).$$

Then after some algebra, we have

$$(I) + (IV) = (-1)\left[(e^{-i\frac{\theta_k}{2}} - e^{i\frac{\theta_k}{2}})^2(e^{-i\frac{\theta_{k'}}{2}} - e^{i\frac{\theta_{k'}}{2}})^2\right]e^{i\frac{\theta_k+\theta_{k'}}{2}},$$

$$(II) + (III) = (-1)\left[(e^{-i\frac{\theta_k}{2}} - e^{i\frac{\theta_k}{2}})^2(e^{-i\frac{\theta_{k'}}{2}} - e^{i\frac{\theta_{k'}}{2}})^2\right]$$
$$\left[1 - (e^{-i\frac{(\theta_k-\theta_{k'})}{2}} + (e^{i\frac{(\theta_k-\theta_{k'})}{2}})\right].$$

Hence we finally find that

$$(I) + (II) + (III) + (IV) = (-1)(e^{-i\frac{\theta_k}{2}} - e^{i\frac{\theta_k}{2}})^2(e^{-i\frac{\theta_{k'}}{2}} - e^{i\frac{\theta_{k'}}{2}})^2$$
$$\left[1 - e^{i\frac{\theta_k-\theta_{k'}}{2}} - e^{-i\frac{\theta_k-\theta_{k'}}{2}} + e^{-i\frac{\theta_k-\theta_{k'}}{2}}\right]$$

and then it is proportional to

$$(I) + (II) + (III) + (IV) \sim 4\left[\sin^2\frac{\theta_k}{2}\right]\left[\sin^2\frac{\theta_{k'}}{2}\right],$$

which is the dominant term, when $k/n \longrightarrow 0$ as $n \to \infty$. (Q.E.D.)

References

Akaike H (1989) Choosing priors and its applications. In: Suzuki (ed) Bayesian statistics and applications. Kunitomo, Tokyo University Press, Y. and N in Japanese

Amemiya T (1985) Advanced econometrics. Blackwell

Anderson TW (2003) An introduction to statistical multivariate analysis, 3rd edn. John-Wiley

Anderson TW (1984) Estimating linear statistical relationships. Ann Stat 12:1–45

Anderson TW, Takemura A (1986) Why do non-invertible moving average occur? J Time Series Anal 7–4:235–254

Durrett R (1991) Probability: theory and examples. Duxbury Press

Engle R, Granger CWJ (1987) Co-integration and error correction. Econometrica 55:251–276

Fuller W (1987) Measurement error models. John-Wiley

Granger CWJ, Newbold P (1977) Forecasting economic time series. Academic Press

Hayashi F (2000) Econometrics. Princeton University Press

Johansen S (1995) Likelihood based inference in cointegrated vector autoregressive models. Oxford UP

Kitagawa G (2021) Introduction to time series modeling with applications in R, 2nd edn. CRC Press

Kunitomo N, Sato S (2017) Trend, seasonality and economic time series: the nonstationary errors-in-variables models. MIMS-RBP-SDS-3, MIMS, Meiji University. http://www.mims.meiji.ac.jp/

Kunitomo N, Sato S, Kurisu D (2018) Separating information maximum likelihood method for high-frequency financial data. Springer

Chapter 5
Frequency Regression and Smoothing for Noisy Non-stationary Multivariate Time Series

Abstract The frequency regression and smoothing method, or the SIML frequency method, is developed based on the non-stationary errors-in-variables model. Many macroeconomic time series contain not only trend, cycle, seasonal, and measurement error components, but also factors such as abrupt changes, trading-day effects, and institutional changes. The frequency regression and smoothing method is an effective tool for such factors in non-stationary time series. The proposed method is simple and applicable to analyzing non-stationary economic time series and to handle seasonal adjustments. Our formulation leads to the asymptotic results on the low-frequency method proposed by Müller and Watson (Econometrica 86-3:775–804, 2018) as a consequence. To illustrate the method, we present empirical example.

5.1 Introduction

This chapter builds upon the SIML filtering or smoothing method based on the non-stationary errors-in-variables model, which was introduced in Chaps. 2 and 3, to estimate hidden states of random variables and handle multiple time series data. Our method is based on the frequency domain analysis of non-stationary time series and it can be applicable to small sample economic data. We develop a linear regression method in the frequency domain for non-stationary multivariate time series.

As we discussed in Chaps. 2 and 3, macroeconomic variables include important factors such as structural break, trading-day effects, and institutional changes in addition to trend, cycle, and seasonal components as well as the measurement errors. Recent (vivid) examples are the macro-effects of pandemic COVID-19 in 2020–2023 and the global financial crisis in 2008–2009. Another feature is the fact that measurement errors in the economic time series play an important role because many macroeconomic data are constructed from various sources including sample surveys from major official statistics, whereas the statistical time series analysis often ignores measurement errors. Further, many official agencies in the world apply the X-12-ARIMA or X-13ARIMA-SEATS programs of the U.S. Census Bureau, which use the univariate Reg-ARIMA model to remove seasonality, as the standard filtering procedure to publish the seasonally adjusted series.[1]

The quarterly GDP series and its major components in Japan, which are the most important data in Japanese macroeconomy, have been constructed since 1994 by ESRI, the Cabinet Office of Japan. In contrast to the macroeconomic data of the United States, both the original and seasonally adjusted data have been published from ESR, the Cabinet Office of Japan. This provides an opportunity to access appropriateness of official seasonal adjustment applied to the original series. Given the limited sample size, it is crucial to use an appropriate statistical procedure to extract information on the trend-cycle, seasonal, and noise (or measurement error) components in a systematic manner from the data.

Because there are many factors in non-stationary time series, there is no statistical method that can deal with them in a systematic and coherent manner which has yet to be developed. The proposed SIML method for the analysis of non-stationary multivariate time series can be applied to handle these factors systematically. It is simple and applicable to several problems when analyzing a non-stationary multivariate economic time series. As earlier studies on economic time series, Granger and Hatanaka (1964) and Brillinger and Hatanaka (1969) introduced the spectral and harmonic analysis of economic time series. Engle (1974) proposed the band spectrum regression for stationary economic time series. Also our work is closely related to the problem of Baxter and King (1999) and Müller and Watson (2018). In particular, our formulation leads to some asymptotic results on the low-frequency method proposed by Müller and Watson (2018) as a consequence.

In Sect. 5.2, we explain the non-stationary errors-in-variables model and the SIML filtering (or smoothing) method. Then, in Sect. 5.3, we introduce the frequency regression method, and as an application, we mention the result obtained by Müller and Watson (2018). In Sect. 5.4, we discuss the regression smoothing method, which is based on SIML smoothing. In Sect. 5.5, we discuss the likelihood function, and in Sect. 5.6, we show an illustrative empirical analysis of the macro consumption of durable goods in Japan. In Sect. 5.7, we provide some concluding remarks. Some details of the mathematical derivations of the theoretical results on frequency regression and the corresponding figures are presented in the Appendix of this chapter.

5.2 Non-stationary Errors-in-Variables Models and SIML Filtering

Let y_{ji} be the i-th observation of the j-th time series at i for $i = 1, \ldots, n; j = 1, \ldots, p$. Let $\mathbf{y}_i = (y_{1i}, \ldots, y_{pi})'$ be a $p \times 1$ vector and $\mathbf{Y}_n = (\mathbf{y}_i') (= (y_{ij}))$ be an $n \times p$ matrix of observations, further let $\mathbf{y}_0$ be the initial $p \times 1$ vector. We consider the statistical time series model with several unobservable components when we have the underlying non-stationary component $\mathbf{x}_i$ $(i = 1, \ldots, n)$, which is an $I(1)$ process,

---

[1] See HP of U.S. Census Bureau, https://www.census.gov/data/software/x13as.html.

while the vector of the noise (or measurement error) component $\mathbf{v}'_i = (v_{1i}, \ldots, v_{pi})$, which may include the seasonal component, is an $I(0)$ process. We use the non-stationary errors-in-variables representation in the additive form

$$\mathbf{y}_i = \mathbf{x}_i + \mathbf{v}_i \quad (i = 1, \ldots, n). \tag{5.1}$$

We assume that $\mathbf{x}_i$ and $\mathbf{v}_i$ $(i = 1, \ldots, n)$ are mutually independent for the simplicity. The non-stationary state variable satisfies

$$\Delta \mathbf{x}_i = (1 - \mathscr{L})\mathbf{x}_i = \mathbf{v}_i^{(x)} \tag{5.2}$$

with the lag operator $\mathscr{L}\mathbf{x}_i = \mathbf{x}_{i-1}$, $\Delta = 1 - \mathscr{L}$. (See (2.11) and (2.12) in Sect. 2.2.2.)

In this chapter, we use the general non-stationary errors-in-variables model that $\mathbf{v}_i^{(x)} = \sum_{j=0}^{\infty} \mathbf{C}_j^{(x)} \mathbf{e}_{i-j}^{(x)}$, where $\mathbf{e}_i^{(x)}$ denotes a sequence of i.i.d. random vectors with $\mathbf{E}(\mathbf{e}_i^{(x)}) = \mathbf{0}$ and $\mathbf{E}(\mathbf{e}_i^{(x)}\mathbf{e}_i^{(x)'}) = \boldsymbol{\Sigma}_e^{(x)}$ (a positive-semi-definite matrix). The $p \times p$ coefficient matrices $\mathbf{C}_j^{(x)}$ $(= c_{kl}^{(x)}(j))$ are absolutely summable and $\|\mathbf{C}_j^{(x)}\| = O(\rho^j)$, where $0 \leq \rho < 1$. The initial value $\mathbf{y}_0$ $(= \mathbf{x}_0)$ is fixed. The stationary noise component $\mathbf{v}_i$ satisfies $\mathbf{v}_i = \sum_{j=0}^{\infty} \mathbf{C}_j^{(v)} \mathbf{e}_{i-j}^{(v)}$, where the $p \times p$ coefficient matrices $\mathbf{C}_j^{(v)}$ are absolutely summable and $\|\mathbf{C}_j^{(v)}\| = O(\rho^j)$, where $0 \leq \rho < 1$ and $\mathbf{e}_i^{(v)}$ represents a sequence of i.i.d. random vectors with $\mathbf{E}(\mathbf{e}_i^{(v)}) = \mathbf{0}$ and $\mathbf{E}(\mathbf{e}_i^{(v)}\mathbf{e}_i^{(v)'}) = \boldsymbol{\Sigma}_e^{(v)}$ (positive definite matrix). For normalization, we use the notation $\mathbf{C}_0^{(v)} = \mathbf{C}_0^{(x)} = \mathbf{I}_p$.

There are several decomposition models of time series based on (5.1)–(5.2). When the state variables of our interest are the trend-cycle components $\mathbf{TC}_i$ ($p \times 1$ vectors), we take $\mathbf{x}_i = \mathbf{TC}_i$ $(i = 1, \ldots, n)$ and $\Delta \mathbf{TC}_i = \mathbf{TC}_i - \mathbf{TC}_{i-1}$ is a stationary process, which has the MA representation. In this case $\mathbf{v}_i$ is the stationary error process including the measurement errors except the trend-cycle components.

An important example is the seasonal adjustment such as X-13ARIMA-SEATS of the U.S. Census Bureau, which usually estimates the seasonal components to construct seasonal adjusted series. When the state variables of our interest are the trend-cycle components $\mathbf{TC}_i$ and the seasonal component $\mathbf{s}_i$ ($p \times 1$ vectors), we may interpret that $\Delta \mathbf{TC}_i = \mathbf{TC}_i - \mathbf{TC}_i$ and $\mathbf{s}_i$ are stationary processes, and $\Delta \mathbf{x}_i = \Delta \mathbf{TC}_i + \mathbf{s}_i$ has the MA representation of (2.11). In this case, we may have $\Delta \mathbf{TC}_i = \sum_{j=0}^{\infty} \mathbf{C}_j^{(TC)} \mathbf{e}_{i-sj}^{(TC)}$ and $\mathbf{s}_i = \sum_{j=0}^{\infty} \mathbf{C}_{sj}^{(s)} \mathbf{e}_{i-sj}^{(s)}$, where the lag operator is defined by $\mathscr{L}^s \mathbf{s}_i = \mathbf{s}_{i-s}$ $(s \geq 2)$, and $\mathbf{e}_i^{(TC)}$ and $\mathbf{e}_i^{(s)}$ represent sequences of i.i.d. random vectors. That is, $\mathbf{E}(\mathbf{e}_i^{(TC)}) = \mathbf{E}(\mathbf{e}_i^{(s)}) = \mathbf{0}$ and $\mathbf{E}(\mathbf{e}_i^{(TC)}\mathbf{e}_i^{(TC)'}) = \boldsymbol{\Sigma}_e^{(TC)}$ (non-negative definite matrix) and $\mathbf{E}(\mathbf{e}_i^{(s)}\mathbf{e}_i^{(s)'}) = \boldsymbol{\Sigma}_e^{(s)}$ (non-negative definite matrix). The $p \times p$ coefficient matrices $\mathbf{C}_j^{(TC)}$ and $\mathbf{C}_j^{(s)}$ are absolutely summable such that $\|\mathbf{C}_j^{(TC)}\| = O(\rho^j)$ and $\|\mathbf{C}_j^{(s)}\| = O(\rho^j)$, where $0 \leq \rho < 1$.

We apply the general filtering procedure based on the $\mathbf{K}_n^*$-transformation and then, it is easy to interpret the role of the elements of the resulting $n \times p$ random matrix $\mathbf{Z}_n$ in the data analysis because they are obtained by the transformation that takes real values in the frequency domain.

Let an $n \times p$ matrix

$$\hat{\mathbf{X}}_n = \mathbf{C}_n \mathbf{P}_n \mathbf{Q}_n \mathbf{P}_n \mathbf{C}_n^{-1} (\mathbf{Y}_n - \bar{\mathbf{Y}}_0), \tag{5.3}$$

where $\mathbf{Z}_n = \mathbf{P}_n \mathbf{C}_n^{-1} (\mathbf{Y}_n - \bar{\mathbf{Y}}_0)$ and $\mathbf{Q}_n$ denotes an $n \times n$ filtering matrix.

5.3   Frequency Regression

In this section, we partition the transformed random variables in the frequency domain and consider a linear regression model based on observations of $q \times p$ matrix $\mathbf{Z}_m^*$ by

$$\mathbf{Z}_m^* = \mathbf{F}_q \mathbf{P}_n \mathbf{C}_n^{-1} (\mathbf{Y}_n - \bar{\mathbf{Y}}_0) = [\mathbf{z}_{1m}^*, \mathbf{Z}_{2m}^*], \tag{5.4}$$

where $\mathbf{F}_q$ denotes a $q \times n$ matrix, $q \, (> p)$ may depend on n as $q = q_n$, $\mathbf{z}_{1m}^*$ is a $q \times 1$ vector, and $\mathbf{Z}_{2m}^*$ is a $q \times (p-1)$ matrix.

There are several interesting examples. Since we consider the case when the rank of $\mathbf{F}_q$ is $p \, (p < q)$, let us investigate this case.

When we have non-stationary time series, we often have trend, cycle, seasonal, and noise components. To handle these components, we can use a more complicated transformation $\mathbf{F}_q$. Further, there are trading-day components, leap year effects, structural changes such as the 2008 financial crisis and the 2020–2022 COVID-19 crisis, and institutional changes such as the consumption tax in Japan. When we use seasonal adjusted data, which are published by official agencies, it is important to handle these effects in meaningful ways.

We consider the simple case when we set $\mathbf{F}_q = \mathbf{J}_m$. We first investigate this case and assume that the rank of $\mathbf{F}_q$ is $p \, (p < q)$. We define $p \times p$ matrices

$$\mathbf{G}_m^* = \frac{1}{m} \mathbf{Z}_m^{*'} \mathbf{Z}_m^*, \quad \mathbf{G}_n = \frac{1}{n} \mathbf{Z}_n' \mathbf{Z}_n, \tag{5.5}$$

where $\mathbf{G}_n$ is constructed by setting $n = m = q$.

Their probability limits as $m = m_n \to \infty \ (n \to \infty, m_n/n \to 0)$

$$\text{plim}_{n\to\infty} \mathbf{G}_m^* = \mathbf{\Sigma}_x, \quad \text{plim}_{n\to\infty} \mathbf{G}_n = \mathbf{\Sigma}_{\Delta y}, \tag{5.6}$$

where

$$\mathbf{\Sigma}_x = \left(\sum_{j=0}^{\infty} \mathbf{C}_j^{(x)} \right) \mathbf{\Sigma}_e^{(x)} \left(\sum_{j=0}^{\infty} \mathbf{C}_j^{(x)'} \right) \quad (= \mathbf{f}_{\Delta x}(0)), \tag{5.7}$$

where $\mathbf{\Sigma}_x$ represents the spectral density matrix of $\Delta \mathbf{x}_i$ at zero frequency. The $p \times p$ matrix $\mathbf{\Sigma}_{\Delta y}$ is the spectral density matrix of $\Delta \mathbf{y}_i$ at zero frequency, which is different from $\boldsymbol{\Sigma}_x$ (the long-run variance-covariance matrix of $\Delta\mathbf{x}_i$) in the errors-in-variables models of (5.1)–(5.2).

Let $\boldsymbol{\Sigma}_v = (\sum_{j=0}^{\infty} \mathbf{C}_j^{(v)}) \boldsymbol{\Sigma}_e^{(v)} (\sum_{j=0}^{\infty} \mathbf{C}_j^{(v)'})$, which is a $p \times p$ positive definite matrix because we assumed that $\boldsymbol{\Sigma}_e^{(v)}$ is positive definite. This assumption on the errors-in-variables models has an important role because we have both the signal and noise terms in the observed time series.

We partition $\mathbf{G}_m^*$, $\boldsymbol{\Sigma}_x$, and $\boldsymbol{\Sigma}_v$ ($p \times p$ matrices) into $(1+k) \times (1+k)$ ($k = p - 1$) submatrices as

$$\mathbf{G}_m^* = \begin{bmatrix} g_{11}^* & \mathbf{g}_{12}^* \\ \mathbf{g}_{21}^* & \mathbf{G}_{22}^* \end{bmatrix}, \quad \boldsymbol{\Sigma}_x = \begin{bmatrix} \sigma_{11}^{(x)} & \sigma_{12}^{(x)} \\ \sigma_{21}^{(x)} & \boldsymbol{\Sigma}_{22}^{(x)} \end{bmatrix}, \quad \boldsymbol{\Sigma}_v = \begin{bmatrix} \sigma_{11}^{(v)} & \sigma_{12}^{(v)} \\ \sigma_{21}^{(v)} & \boldsymbol{\Sigma}_{22}^{(v)} \end{bmatrix}. \tag{5.8}$$

Then, we will investigate statistical properties of the least squares estimator in the frequency domain

$$\hat{\boldsymbol{\beta}}_m = \mathbf{G}_{22}^{*-1} \mathbf{g}_{21}^*, \tag{5.9}$$

which is an estimator of vector $\boldsymbol{\beta}_m = \boldsymbol{\Sigma}_{22}^{(x)-1} \sigma_{21}^{(x)}$ under the assumption that the inverse matrices of $\mathbf{G}_{22}^*$ and $\boldsymbol{\Sigma}_{22}^{(x)}$ exist. (We need to assume that $\boldsymbol{\Sigma}_{22}^{(x)}$ has a full rank.)

We write

$$\hat{\boldsymbol{\beta}}_m - \boldsymbol{\beta} = [\mathbf{Z}_{2m}^{*'}\mathbf{Z}_{2m}^*]^{-1}\mathbf{Z}_{2m}^{*'}\mathbf{Z}_m^* \begin{pmatrix} 1 \\ -\boldsymbol{\beta} \end{pmatrix}, \tag{5.10}$$

where we partitioned $\mathbf{Z}_m^*$ into $q = m \times (1+k)$ submatrices $\mathbf{Z}_m^* = (\mathbf{z}_{1m}^*, \mathbf{Z}_{2m}^*)$.

Then we have the next result on the consistency and asymptotic normality of the frequency least squares estimator and the proof is presented in the Appendix of this chapter.

Theorem 5.1 *Let* $m_n = [n^\alpha]$, $m = [m_n]$ *and* $m \to \infty$ *(as* $n \to \infty$*). In (5.1) and (5.2), assume that the fourth-order moments of* $\mathbf{e}_i^{(x)}$ *and* $\mathbf{e}_i^{(v)}$ *are bounded.*

(i) *For* $0 < \alpha < 1$, $\mathbf{G}_m^*$ *is a consistent estimator of* $\boldsymbol{\Sigma}_x$ *as* $n \to \infty$.

(ii) *Assume that the rank of* $\boldsymbol{\Sigma}_{22}^{(x)}$ *is* $p - 1$ *and* $0 < \alpha < 0.8$. *Then when* $m \to \infty$ *(n* $\to \infty$*),* $\sqrt{m_n}[\hat{\boldsymbol{\beta}}_m - \boldsymbol{\beta}]$ *is asymptotically and normally distributed as* $N(\mathbf{0}, \sigma_{11.2}\boldsymbol{\Sigma}_{22}^{(x)-1})$ *and* $\sigma_{11.2}^{(x)} = \sigma_{11}^{(x)} - \sigma_{12}^{(x)}\boldsymbol{\Sigma}_{22}^{(x)-1}\sigma_{21}^{(x)}$.

Then, we can rewrite $\mathbf{u}_m = \mathbf{z}_{1m}^* - \mathbf{Z}_{2m}^*\boldsymbol{\beta}$, that is,

$$\mathbf{z}_{1m}^* = \mathbf{Z}_{2m}^*\boldsymbol{\beta} + \mathbf{u}_m \tag{5.11}$$

and $\mathscr{E}[\mathbf{u}_m] = \mathbf{0}$. This is a linear regression equation in the frequency domain; however, the error term of $\mathbf{u}_m$ has a specific form of heteroscedasticity.

Theorem 5.1 is valid when we estimate the covariance matrix $\boldsymbol{\Sigma}_x$ of the hidden state variables, which is different from the observed covariance of the differenced data $\boldsymbol{\Sigma}_{\Delta y}$. We can delete the effects of noisy parts of non-stationary time series by using $\mathbf{G}_m$ instead of $\mathbf{G}_n$ with the condition $m_n/n \to 0$ and $m_n \to \infty$. By the condition $0 < \alpha < 0.8$, we can recover the asymptotic normality of the least squares estimator without noises.

One direct application of Theorem 5.1 is Müller and Watson (2018), who proposed the *so-called* long-run co-variability of macroeconomic time series when $p = 2$. (See discussion in Sect. 3.4.1.) They investigated many non-stationary time series using their method and obtained some interesting findings. We can interpret their method as the relationships among long-run trends in our framework when $p = 2$. Let 2×2 matrices $\boldsymbol{\Sigma}_e^{(x)} = (\sigma_{ij}^{(x)})$; then, we define the regression coefficient $\boldsymbol{\beta} = [\sigma_{22}^{(x)}]^{-1}\sigma_{21}^{(x)}$ under the assumption that $\sigma_{22}^{(x)} (= \boldsymbol{\Sigma}_{22}^{(x)}) > 0$.

Further, let $\boldsymbol{G}_m^* = (\hat{g}_{ij}^{(x)})$, and an $n \times 2$ matrix

$$(\mathbf{a}_{1n}, \mathbf{a}_{2n}) = \mathbf{C}_n^{-1}(\mathbf{Y}_n - \mathbf{Y}_0). \tag{5.12}$$

For estimating $\boldsymbol{\beta}$, we define the estimated regression coefficient as

$$\hat{\boldsymbol{\beta}} = [\hat{g}_{22}^{(x)}]^{-1}\hat{g}_{21}^{(x)} = [\mathbf{a}_{2n}'\mathbf{P}_n\mathbf{J}_m\mathbf{J}_m'\mathbf{P}_n\mathbf{a}_{2n}]^{-1}[\mathbf{a}_{2n}'\mathbf{P}_n\mathbf{J}_m\mathbf{J}_m'\mathbf{P}_n\mathbf{a}_{1n}]. \tag{5.13}$$

This quantity can be interpreted as the least squares slope of the transformed vector from $\mathbf{y}_{1n}$ on the transformed vector from $\mathbf{y}_{2n}$ for a $n \times 2$ matrix $\mathbf{Y}_n = (\mathbf{y}_{1n}, \mathbf{y}_{2n})$; that is, essentially the same as the estimation method proposed by Müller and Watson (2018).[2] However, there is an important difference between the SIML method and their method, that is, we consider the situation when $m = [m_n] \to \infty$ as $n \to \infty$ while $m_n/n \to 0$. This is a natural framework for the asymptotic theory on their method.

We fix m, which is independent from n and we investigate the case when $\Delta \mathbf{x}_i$ and $\mathbf{v}_i$ $(i = 1, \ldots, n)$ are mutually independent with $\mathbf{y}_0 = \mathbf{x}_0 = \mathbf{0}$ for the simplicity. Because of (5.1) and (5.2), we find that $\boldsymbol{\Sigma}_{\Delta y} = \boldsymbol{\Sigma}_x + 2\boldsymbol{\Sigma}_v$. Then, as $n \to \infty$,

$$\hat{\boldsymbol{\beta}} \xrightarrow{p} [\boldsymbol{\Sigma}_{22}^{(x)} + 2\boldsymbol{\Sigma}_{22}^{(v)}]^{-1}[\sigma_{21}^{(x)} + 2\sigma_{21}^{(v)}]. \tag{5.14}$$

This corresponds to the fact that the least squares estimator is not consistent when the sample size is large in the classical errors-in-variables models for i.i.d. observations. (See Anderson (1984) in detail.) When $\boldsymbol{\Sigma}_v$ is relatively small, the probability limit of $\hat{\boldsymbol{\beta}}$ is close to $\boldsymbol{\beta}$ because the error terms $\mathbf{v}_i$ are negligible. From Theorem 5.1 (and Theorem 5.3 in the Appendix of this chapter), we obtain the following result.

Corollary 5.1 *When $p = 2$, we assume that $\boldsymbol{\Sigma}_e^{(x)}$ is positive-semi-definite, $\boldsymbol{\Sigma}_e^{(v)}$ is positive definite, and the fourth-order moments of $\mathbf{e}_i^{(x)}$ and $\mathbf{e}_i^{(v)}$ $(i = 1, \ldots, n)$ are bounded.*

---

[2] In their notation, m corresponds to q, which is fixed. They did use (differenced) stationary data, and thus, we could interpret that they calculated the linear regression from the filtered data $\hat{\mathbf{X}}_n^* = \mathbf{P}_n'\mathbf{J}_m'\mathbf{J}_m\mathbf{P}_n\mathbf{C}_n^{-1}(\mathbf{Y}_n - \mathbf{Y}_0)$ as a modification of (2.11) in our notation.

(i) *Fix m, which is independent of n. Then $\hat{\beta}$ in (3.9) is not consistent when $n \to \infty$.*

(ii) *Set $m_n = n^\alpha$ and $0 < \alpha < 1$ and construct $\hat{\beta}_m$ in (3.6). Then, as $n \to \infty$,*
$$\hat{\beta}_m - \beta \xrightarrow{p} 0.$$

(iii) *Set $m_n = n^\alpha$ and $0 < \alpha < 0.8$, then, as $n \to \infty$, $\sqrt{m_n}[\hat{\beta}_m - \beta]$ is asymptotically normal with $N(0, \sigma_{11.2}^{(x)})$.*

This is an extension of Proposition 3.3 in Chap. 3. Since our method can be generalized to other situations beyond the trend-cycle components of non-stationary time series, it is a generalization of Müller and Watson (2018). For instance, it is rather straightforward to incorporate the regression effects of dummy variables in trend relations such as structural breaks and the seasonal frequency parts.

5.4 Regression Smoothing

When we have noisy non-stationary time series, we often need to remove the seasonality and/or low-frequency component. However, for applications in official statistics, we need to construct the seasonally adjusted data after removing additional effects such as trading-day components including the leap year effect, structural changes such as the 2008 financial crisis, and institutional changes such as the introduction of consumption tax in Japan. These effects can be defined in deterministic ways.

Let the observed vector times series $\mathbf{y}_i$ be decomposed as

$$\mathbf{y}_i = \mathbf{x}_i + \mathbf{SCO}_i + \mathbf{v}_i \quad (i = 1, \ldots, n), \tag{5.15}$$

and $\mathbf{SCO}_i = \mathbf{SC}_i + \mathbf{O}_i$, where $\mathbf{x}_i$ denotes the trend-cycle component, $\mathbf{SC}_i$ denotes the structural break component, $\mathbf{v}_i$ denotes the noise component, and $\mathbf{O}_i$ represents the outlier component.

In this section we consider the case where $\mathbf{SC}_i$ and $\mathbf{O}_i$ can be expressed as $\mathbf{SC}_i + \mathbf{O}_i = \mathbf{SCO}_i(w)$, where $\mathbf{w}$ denotes the set of instrumental variables or exogenous variables. If these terms can be expressed as linear relationships, we write

$$\mathbf{y}_i = \mathbf{B}'\mathbf{w}_i + \mathbf{u}_i \quad (i = 1, \ldots, n), \tag{5.16}$$

where $\mathbf{B}'$ denotes a $p \times r$ matrix, $\mathbf{w}_i$ denotes a $r \times 1$ vector of instrumental variables, and $\mathbf{z}_i$ and $\mathbf{u}_i = \mathbf{x}_i + \mathbf{v}_i$ represent a sequence of I(1) process. Hence, the model is a multivariate regression model when the error terms are $I(1)$ process with stationary noise term and seasonal terms. We incorporate extraneous information such as dummy variables to extract or delete some components from the observed time series based on (5.1).

To find the regression and smoothing procedure of trend and seasonal components, we use the K_n^*-transformation of data and rewrite (5.16) as

$$\mathbf{Y}_n^* = \mathbf{W}_n^*\mathbf{B} + \mathbf{U}_n^*, \tag{5.17}$$

where $\mathbf{Y}_n^* = \mathbf{P}_n\mathbf{C}_n^{-1}(\mathbf{Y}_n - \mathbf{Y}_0)$ and $\mathbf{W}_n^* = \mathbf{P}_n\mathbf{C}_n^{-1}\mathbf{W}_n$ ($\mathbf{W}_n = (\mathbf{w}_t')$) represent $n \times p$ and $n \times r$ matrices of the explained variables and explanatory variables, respectively, and $\mathbf{U}_n^* = \mathbf{P}_n\mathbf{C}_n^{-1}\mathbf{U}_n$ and $\mathbf{U}_n = (\mathbf{u}_i')$ are $n \times p$ disturbance matrices. (We fix the initial condition $\mathbf{y}_0$ ($= \mathbf{x}_0$) and the state variables $\mathbf{x}_i^* = \mathbf{x}_i - \mathbf{x}_0$.)

As a consequence of the K_n^*-transformation, we have the disturbance terms in (5.3), that are stationary processes.

Because (5.17) is a linear regression equation, it is possible to apply Theorem 5.1 by defining a $(p + r) \times 1$ vector

$$y_i^* = \begin{bmatrix} \mathbf{y}_i \\ \mathbf{w}_i \end{bmatrix}.$$

Then we can estimate the regression coefficients and calculate the residuals from the regression equations. When vectors $\mathbf{w}_i$ ($i = 1, \ldots, n$) are deterministic, we assume that

$$\lim_{m \to \infty} \frac{1}{m}\mathbf{W}_m^{*'}\mathbf{W}_m^* = \boldsymbol{\Sigma}_{w^*}, \tag{5.18}$$

where $\boldsymbol{\Sigma}_{w^*}$ denotes a positive definite matrix and $\mathbf{W}_m^* = \mathbf{J}_m\mathbf{P}_n\mathbf{C}_n^{-1}\mathbf{W}_n$ represents an $m \times r$ matrix.

When the $r \times 1$ instrumental variables $\mathbf{w}_i$ ($i = 1, \ldots, n$) are exogenous or deterministic, we have the following result from Theorem 5.1 and the proof is in the Appendix of this chapter.

Theorem 5.2 *In (5.1), (5.2), and (5.15), assume that the fourth-order moments of $\mathbf{e}_i^{(x)}$ and $\mathbf{e}_i^{(v)}$ are bounded. Let $\mathbf{Y}_m^* = \mathbf{W}_m^*\mathbf{B} + \mathbf{U}_m^*$ and $\mathbf{U}_m^* = \mathbf{J}_m\mathbf{U}_n^*$, where $\mathbf{Y}_m^* = \mathbf{J}_m\mathbf{P}_n\mathbf{C}_n^{-1}\mathbf{Y}_n$ and $\mathbf{W}_m^* = \mathbf{J}_m\mathbf{P}_n\mathbf{C}_n^{-1}\mathbf{W}_n$. We also assume the nonsingularity condition (5.18) and $\mathbf{w}_i$ ($i = 1, \ldots, n$) is exogenous or deterministic function.*

We denote $\hat{\mathbf{B}}_m = (\mathbf{W}_m^{*'}\mathbf{W}_m^*)^{-1}\mathbf{W}_m^{*'}\mathbf{Y}_m^*$ is the least squares estimator of $\mathbf{Y}_m^*$ on $\mathbf{W}_m^*$. Let $m_n = n^\alpha$ ($0 < \alpha < 0.8$) and $m = [m_n]$. Then, as $n \longrightarrow \infty$, we have the asymptotic normality

$$\sqrt{m_n}[\hat{\mathbf{B}}_m - \mathbf{B}] \xrightarrow{w} N(\mathbf{0}, \boldsymbol{\Sigma}_{w^*}^{-1} \otimes \boldsymbol{\Sigma}_x). \tag{5.19}$$

Define the general transformed instrumental variables

$$\hat{\mathbf{W}}_n = \mathbf{J}_W\mathbf{P}_n\mathbf{C}_n^{-1}\mathbf{W}_n , \tag{5.20}$$

where $\mathbf{J}_W$ represents a $q \times n$ choice matrix, and we denote the idempotent matrix ($q \times q$ matrix)

$$\mathbf{Q}_W = \hat{\mathbf{W}}_n(\hat{\mathbf{W}}_n'\hat{\mathbf{W}}_n)^{-1}\hat{\mathbf{W}}_n'. \tag{5.21}$$

(We assume the condition $q > r$.)

We utilize the regression information on smoothing by utilizing the projection matrix $\mathbf{Q}_W$ to construct

$$\hat{\mathbf{X}}_n^{(w)} = \mathbf{C}_n \mathbf{P}_n \mathbf{J}_W' \mathbf{Q}_W \mathbf{J}_W \mathbf{P}_n \mathbf{C}_n^{-1} (\mathbf{Y}_n - \bar{\mathbf{Y}}_0). \tag{5.22}$$

There are several possibilities to how we incorporate the extraneous information in the smoothing procedure. It is reasonable to consider the case when $\mathbf{Q}_W$ is an idempotent matrix such as $\mathbf{Q}_W^2 = \mathbf{Q}_W$. In this chapter, we use two alternative smoothing procedures: *Type-I* and *Type-II*. Type-I smoothing may be appropriate for change-point smoothing in the trend-cycle component and Type-II smoothing may be appropriate for seasonal adjustments and outlier detection in the noise component.

Type-I Smoothing

Type-I is based on Example 3.1 of Sect. 3.1. The (trend-cycle) regression part of $\mathbf{Y}_n$ is (5.1) when we take $\mathbf{J}_W = \mathbf{J}_m = (\mathbf{I}_m, \mathbf{O})$ ($\hat{\mathbf{W}}_n$ is an $m \times n$ matrix, and $\mathbf{J}_m$ is an $m \times n$ matrix), and an $m \times m$ matrix

$$\mathbf{Q}_m^{(1)} = \hat{\mathbf{W}}_n (\hat{\mathbf{W}}_n' \hat{\mathbf{W}}_n)^{-1} \hat{\mathbf{W}}_n'. \tag{5.23}$$

If we want to remove the regression effects and use only the trend-cycle part, we need to take $\mathbf{J}_W = \mathbf{J}_m$ and

$$\mathbf{Q}_m^{(2)} = \mathbf{I}_m - \mathbf{Q}_m^{(1)} = \mathbf{I}_m - \hat{\mathbf{W}}_n (\hat{\mathbf{W}}_n' \hat{\mathbf{W}}_n)^{-1} \hat{\mathbf{W}}_n'. \tag{5.24}$$

Then we have the decomposition

$$\begin{aligned} \hat{\mathbf{X}}_n^{(m)} &= \mathbf{C}_n \mathbf{P}_n \mathbf{J}_m' \mathbf{J}_m \mathbf{P}_n \mathbf{C}_n^{-1} (\mathbf{Y}_n - \bar{\mathbf{Y}}_0) \\ &= \mathbf{C}_n \mathbf{P}_n \mathbf{J}_m' [\mathbf{Q}_n^{(1)} + \mathbf{Q}_n^{(2)}] \mathbf{J}_m \mathbf{P}_n \mathbf{C}_n^{-1} (\mathbf{Y}_n - \bar{\mathbf{Y}}_0). \end{aligned} \tag{5.25}$$

In this case, we have the property $\mathbf{Q}_n^2 = \mathbf{Q}_n = \mathbf{Q}_n^{(1)} + \mathbf{Q}_n^{(2)} = \mathbf{I}_m$, and we have the decomposition of the trend-cycle part and the regression part. There is a simple interpretation of this smoothing because we use only the regression part at m low frequencies.

Type-II Smoothing

Type-II smoothing is based on Example 3.2 of Sect. 3.1. When we need to estimate not only the trend component, but also the noise component, it is important to estimate structural changes and outlier components consistently.

For the seasonal adjustment of time series, we need to estimate the seasonal component for obtaining the seasonally adjusted series, and it is a generalization of Example 3.2 of Sect. 3.1. Thus, we construct an $q \times n$ choice matrix $\mathbf{F}_q$ such that the seasonal components can be removed in their frequencies.

When $s = 4$, we want to remove the data with frequencies around $\lambda_s = 1/4, 1/2$ (1/2 corresponds to the cycle of 2 quarters and 1/4 corresponds to the cycle of 4 quarters). However, we cannot distinguish the 4 quarters cycle from the 2 quarters cycle by using quarterly observations. We set $m_1 = [2n/s]$, and an $(n - 2h - 1) \times n$ choice matrix and an $(n - 3h - 2) \times (n - 2h - 1)$ choice matrix as

$$\mathbf{J}_1^Q = \begin{bmatrix} \mathbf{I}_{m_1-(h+1)} & \mathbf{O} & \mathbf{O} \\ \mathbf{O} & \mathbf{O} & \mathbf{I}_{n-m_1-h} \end{bmatrix}, \ \mathbf{J}_2^Q = \begin{bmatrix} \mathbf{I}_{n-3h-2}, & \mathbf{O} \end{bmatrix}. \tag{5.26}$$

Then we take a $q \times n$ matrix

$$\mathbf{F}_q^Q = \mathbf{J}_2^Q \mathbf{J}_1^Q \tag{5.27}$$

with a small positive integer $h > 0$.

When $s = 12$, we need a more complicated transformation to remove seasonality because we cannot distinguish the 12 months cycle from the 6, 4, 3, 2.4, and the 2 months cycles using monthly observations with frequencies around $\lambda_s = 1/12, 2/12, 3/12, 4/12, 5/12, 6/12$. We set $m_i = i[2n/s]$ and take $(n - i(2h + 1)) \times (n - (i - 1)(2h + 1))$ choice matrices $(i = 1, \ldots, 5)$ and an $(n - 5(2h + 1) - (h + 1)) \times (n - 5(2h + 1))$ choice matrix such that

$$\mathbf{J}_i^M = \begin{bmatrix} \mathbf{I}_{m_i-(i-1)(2h+1)-(h+1)} & \mathbf{O} & \mathbf{O} \\ \mathbf{O} & \mathbf{O} & \mathbf{I}_{n-m_i-h} \end{bmatrix}, \ \mathbf{J}_6^M = \begin{bmatrix} \mathbf{I}_{n-11h-6}, & \mathbf{O} \end{bmatrix}, \tag{5.28}$$

with a small positive integer $h > 0$. To remove the data with seasonal frequencies around λ_{js} $(j = 2, 3, 4, 5)$ using $\mathbf{J}_j^M$ $(j = 1, \ldots, 6)$, we set a $q \times n$ matrix

$$\mathbf{F}_q^M = \prod_{j=1}^{6} \mathbf{J}_{7-j}^M. \tag{5.29}$$

More generally, when we have information of the instrumental variables $\mathbf{W}_n$, we can incorporate the estimated coefficient by regressing

$$\mathbf{Y}_m^* = \mathbf{F}_q \mathbf{P}_n \mathbf{C}_n^{-1} (\mathbf{Y}_n - \bar{\mathbf{Y}}_0) \tag{5.30}$$

to

$$\mathbf{W}_m^* = \mathbf{F}_q \mathbf{P}_n \mathbf{C}_n^{-1} (\mathbf{W}_n - \bar{\mathbf{W}}_0), \tag{5.31}$$

where $\mathbf{F}_q$ (a $q \times n$ matrix) is either $\mathbf{F}_q^Q$ or $\mathbf{F}_q^M$.

Type-II smoothing is defined by

$$\mathbf{Q}_n^{(3)} = \mathbf{W}_n^* (\mathbf{W}_n^{*'} \mathbf{W}_n^*)^{-1} \mathbf{W}_n^{*'} \tag{5.32}$$

and

$$\mathbf{Q}_n^{(4)} = \mathbf{F}_q \mathbf{F}_q' - \mathbf{Q}_n^{(2)}. \tag{5.33}$$

Then, we have the decomposition

$$\hat{\mathbf{X}}_n^{(F)} = \mathbf{C}_n \mathbf{P}_n \mathbf{F}_q' \mathbf{F}_q \mathbf{P}_n \mathbf{C}_n^{-1} (\mathbf{Y}_n - \bar{\mathbf{Y}}_0) \qquad (5.34)$$
$$= \mathbf{C}_n \mathbf{P}_n \mathbf{F}_q' [\mathbf{Q}_n^{(3)} + \mathbf{Q}_n^{(4)}] \mathbf{F}_q \mathbf{P}_n \mathbf{C}_n^{-1} (\mathbf{Y}_n - \bar{\mathbf{Y}}_0).$$

In this case, we have the decomposition $\mathbf{Q}_n^{(3)} + \mathbf{Q}_n^{(4)} = \mathbf{F}_q \mathbf{F}_q' = \mathbf{I}_q$ and the corresponding decomposition of the trend-cycle and regression parts.

Examples of Dummy Variables

There are some examples of outlier and trend dummies.

For non-stationary time series, we should be careful about normalization because there can be significant effects on smoothing. Although there are many other possible dummy variables, we provide some examples that have been used in official data handling such as official seasonal adjustment. In the X-12-ARIMA and X-13ARIMA-SEATS programs, for instance, the Reg-ARIMA modeling uses the following dummy variables. See Census Bureau (2020) for the details of the X-12ARIMA-SEATS program and the list of variables in the Reg-ARIMA modeling.

We give the next list because of an illustration, which will be used in an empirical analysis in Sect. 5.6 below. Let w_s $(s = 1, \ldots, n)$ be the dummy variable.

Example 5.1 The level shift (LS) variable can be defined as $w_s = 0$ if $s < t$ and $w_t = 1$ if $s \geq t$ for $s = 1, \ldots, n$. This can be handled by Type-I smoothing.

Example 5.2 The outlier variable can be defined as $w_s = 1$ if $s = t$ and $w_t = 0$ if $s \neq t$ for $s = 1, \ldots, n$. This variable is often called additive outlier (AO).

Example 5.3 The ramp-dummy variable can be defined by $w_s = 1$ if $s < t_0$, $w_s = 1 - (t - t_0)/(t_1 - t_0)$ if $t_0 \leq t \leq t_1$, and $w_t = 0$ if $s \geq t_1$.

Example 5.4 The double ramp-dummy variable can be defined by $w_s = 1$ if $s < t_0$, $w_s = 1 - (t - t_0)/(t_1 - t_0)$ if $t_0 \leq t \leq t_1$, $w_s = (t - t_1)/(t_2 - t_1)$ if $t_1 \leq t \leq t_2$, and $w_t = c$ if $s \geq t_2$.

5.5 Frequency Domain Analysis and Likelihood

We consider the additive decomposition model $\mathbf{y}_i = \mathbf{x}_i + \mathbf{v}_i$ $(i = 1, \ldots, n)$ of (5.1) and (5.2) in the time domain and give an interpretation on the consequence of the transformation of observation vectors by $\mathbf{K}_n^*$ in (2.6)–(2.9). The transformed random variables $\mathbf{z}_k^{(n)}$ $(k = 1, \ldots, n)$ have a particular structure in the frequency domain. For the resulting simplicity, we take a positive integer m $(= m_n)$.

Let $\mathbf{f}_{\Delta x}(\lambda)$ and $\mathbf{f}_v(\lambda)$ be the spectral density $(p \times p)$ matrices of $\Delta \mathbf{x}_i$ and $\mathbf{v}_i$ $(i = 1, \ldots, n)$, respectively, which are given by (2.14) and (2.15), respectively.

Then, the $p \times p$ spectral density matrix of the transformed vector process, which is observable, and the spectral density of the difference series $\Delta \mathbf{y}_i \ (= \mathbf{y}_i - \mathbf{y}_{i-1})$ can be represented as $\mathbf{f}_{\Delta y}(\lambda) = \mathbf{f}_{\Delta x}(\lambda) + (1 - e^{2\pi i \lambda}) f_v(\lambda)(1 - e^{-2\pi i \lambda})$. We denote the long-run variance-covariance matrix of trends and stationary components for $g, h = 1, \ldots, p$ as $\boldsymbol{\Sigma}_e^{(x)} = \mathbf{f}_{\Delta x}(0) \ (= (\sigma_{gh}^{(x)})), \ \boldsymbol{\Sigma}_e^{(v)} = f_v(0) \ = (\sigma_{gh}^{(v)})$.

Then, we find that $\mathbf{f}_{\Delta y}(0) = \mathbf{f}_{\Delta x}(0)$ at the frequency $\lambda = 0$ and we can ignore the effects of stationary noise terms in (5.1) and (5.2) if we use only the information around the zero frequency from data. That is, the information of the non-stationary trend parts is separated from the information of the stationary noise parts in the frequency domain.

Since the spectral matrices in (2.14)–(2.16) are complex-valued, we symmetrize the spectral density matrices. Then, it is possible to relate the complex-valued spectral matrices to the real-valued random vectors and their likelihood function. By reconsidering the relationship among the continuous-valued discrete time series and the spectral densities, it is possible to interpret the filtered parts and smoothing parts. It has been sometimes neglected in the (standard) statistical time series analysis.

Let $f_v^{(SR)}(\lambda_k)$ and $f_{\Delta x}^{(SR)}(\lambda_k)$ be the symmetrized $p \times p$ spectral matrices of $\mathbf{v}_i$ and $\Delta \mathbf{x}_i$ at $\lambda_k \ (= (k - \frac{1}{2})/(2n + 1))$ for $k = 1, \ldots, n$, that is, $f_v^{(SR)}(\lambda_k) = (1/2)[f_v^{(SR)}(\lambda_k) + \bar{f}_v^{(SR)}(\lambda_k)]$, and $f_{\Delta x}^{(SR)}(\lambda_k) = (1/2)[f_{\Delta x}^{(SR)}(\lambda_k) + \bar{f}_{\Delta x}^{(SR)}(\lambda_k)]$. We denote the $n \times p$ matrix $\mathbf{Z}_n = (\mathbf{z}_k^{(n)}(\lambda_k^{(n)})')$ and $\lambda_k^{(n)} = (k - 1/2)/(2n + 1) \ (k = 1, \ldots, n)$, where $\lambda_k^{(n)}$ corresponds to the frequency of $\mathbf{z}_k^{(n)}$. Then, the transformed random variables are asymptotically orthogonal (or uncorrelated) and the orthogonal processes are approximately distributed as Gaussian distributions when n is large.

If we substitute λ_k into $f_{\Delta y}^{(SR)}(\lambda)$, we find that the variance-covariance matrix of $\mathbf{z}_k^{(n)}$ at λ_k is approximately given by $\mathbf{f}_{\Delta y}^{(SR)}(\lambda_k) = \mathbf{f}_{\Delta x}^{(SR)}(\lambda_k) + a_{kn}^* \mathbf{f}_s^{(SR)}(\lambda_k)$ because $\|1 - e_k^{2\pi i \lambda}\|^2 = 2 - 2\cos(2\pi \lambda_k) = a_{kn}^*$ for $k = 1, \ldots, n$.

Given the initial condition $\mathbf{y}_0$, the (-2) times the conditional log-likelihood function in (5.1)–(5.2) can be approximated except a constant term by

$$l_n = \sum_{k=1}^{n} \log |a_{kn}^* f_v^{(SR)}(\lambda_k) + f_{\Delta x}^{(SR)}(\lambda_k)| \tag{5.35}$$

$$+ \sum_{k=1}^{n} \mathbf{z}_k' \left[a_{kn}^* f_v^{(SR)}(\lambda_k) + f_{\Delta x}^{(SR)}(\lambda_k) \right]^{-1} \mathbf{z}_k$$

provided that $a_{kn}^* f_v^{(SR)}(\lambda_k) + f_{\Delta x}^{(SR)}(\lambda_k)$ are positive definite (a.e.).

In particular, we consider the case when $\Delta \mathbf{x}_i$ and $\mathbf{v}_i$ are a sequence of independent random vectors, then we have $\boldsymbol{\Sigma}_e^{(x)} = f_{\Delta x}^{(SR)}(\lambda_k)$ and $\boldsymbol{\Sigma}_e^{(v)} = f_v^{(SR)}(\lambda_k)$ for $k = 1, \ldots, n$. This corresponds to the case when

$$l_n^* = \sum_{k=1}^{n} \log |a_{kn}^* \boldsymbol{\Sigma}_e^{(v)} + \boldsymbol{\Sigma}_e^{(x)}| + \sum_{k=1}^{n} \mathbf{z}_k' \left[a_{kn}^* \boldsymbol{\Sigma}_e^{(v)} + \boldsymbol{\Sigma}_e^{(x)} \right]^{-1} \mathbf{z}_k, \tag{5.36}$$

provided that $a_{kn}^* \mathbf{\Sigma}_e^{(v)} + \mathbf{\Sigma}_e^{(x)}$ are positive definite (a.e.).

Furthermore, if we we take $k = 1, \ldots, m_n$ such that $m_n/n \to 0$ and $m_n \to \infty$ as $n \to \infty$, we have $a_{kn}^* \to 0$ $(k = 1, \ldots, m_n)$. In this situation, the first m_n terms of (3.37) can be regarded as the log-likelihood function in the low-frequency parts of time series, which is given as

$$
l_{1n}^{**} = \sum_{k=1}^{m} \log |\mathbf{\Sigma}_e^{(x)}| + \sum_{k=1}^{m} \mathbf{z}_k' \left[\mathbf{\Sigma}_e^{(x)} \right]^{-1} \mathbf{z}_k, \tag{5.37}
$$

provided that $\mathbf{\Sigma}_e^{(x)}$ is positive definite.

The last representation corresponds to the Gaussian log-likelihood function based on m mutually independent observations in the statistical multivariate analysis. (Fuller (1987) and Anderson (2003), for instance.)

In this way, we can separate the likelihood information into different parts of time series components in the SIML (separate information maximum likelihood) method. The SIML estimator of $\mathbf{\Sigma}_e^{(x)}$ is constructed by using $\mathbf{z}_k$ $(k = 1, \ldots, m_n)$, and we have some desirable asymptotic properties for $\mathbf{\Sigma}_e^{(x)}$.

When we have some dummy variables $\mathbf{W}_n$, we need to assume that they are independent of other noise, cycle, seasonal, and trend components. There can be several ways to handle explanatory variables as explained as Type-I and Type-II in Sect. 5.4, but we explain a typical case. Given the initial condition and the information set of explanatory variables $\mathbf{W}_n$, (-2) times the conditional log-likelihood can be approximated as (5.36) by $\mathbf{z}_k^*(w) = \mathbf{y}_k^* - \mathbf{B}' \mathbf{w}_k^*$, where $\mathbf{B}$ is the parameter matrix, $\mathbf{y}_k^*$ and $\mathbf{w}_k^*$ are the transformed explained variables and explanatory variables using $\mathbf{K}_n^*$-transformation from the observed $\mathbf{y}_i$ and $\mathbf{w}_i$ $(i = 1, \ldots, n)$, respectively.

When we use the explanatory variables $\mathbf{W}_n$, we can estimate the unknown matrix $\mathbf{B}$ by Theorem 5.1 consistently. Let $\hat{\mathbf{B}}$ be the SIML estimator and $\mathbf{z}_k^* = \mathbf{y}_k^* - \hat{\mathbf{B}}' \mathbf{w}_k^*$ $(k = 1, \ldots, n)$, which depends on $\mathbf{W}_n$ and denote $\mathbf{z}_k^*(w)$ $(k = 1, \ldots, n)$. To estimate $\mathbf{\Sigma}_x$ when there are explanatory variables, for instance, it is reasonable to use

$$
\mathbf{G}_m^*(w) = \frac{1}{m} \sum_{k=1}^{m} \mathbf{z}_k^*(w) \mathbf{z}_k^{*'}(w) \tag{5.38}
$$

because it is consistent and has the asymptotic normality if we take $m = [m_n]$ such that $m_n/n \to 0$, $m_n \to \infty$, and $n \to \infty$.

There are two remarks on the likelihood function of time series data in the above discussion.

First, the ML estimation of unknown parameters in the non-stationary errors-in-variables models may have some difficulty when $p > 1$ without some restrictions of the parameter space. It is because the exact likelihood function can have a peculiar shape even when the observations are the sequence of independent random variables. There could be more complications when we have multivariate time series data with random walks, seasonal components, autocorrelations, and measurement errors. (See Chap. 4.)

Second, the likelihood functions in this section can be related to the classical topic on the *Wittle-type* likelihood function for stationary time series in the literature, which does not depend on the Gaussian distributions for underlying noise distributions in multivariate stationary processes. The maximum of Wittle-type likelihood has been called the quasi-maximum likelihood (QML) estimation, which is discussed and applied by Hosoya (1997), for instance.

5.6 An Empirical Example of Macro Consumption of Durable Goods

We use the official macro consumption data of durable goods in Japan from 1994Q1 to 2019Q4 to illustrate the regression smoothing method.[3] In order to construct the seasonal adjusted data, we need to estimate the seasonal factor before making seasonal adjustments. However, it is first necessary to estimate the trend and noise components to estimate the seasonal component from the original quarterly time series at the same time. The traditional X-13ARIMA-SEATS program of the U.S. Census Bureau employs a rather intricate procedure to do this by utilizing the moving average filters repeatedly.

In our analysis, we applied the SIML smoothing procedure with $m = 29$ (Type-1 smoothing in Sect. 5.4) and $h = 2$ (Type-2 smoothing), which yield the minimum numbers of AIC. Therefore, at each seasonal frequency 1/4 and 1/2, we have chosen 5 and 3 frequency data points in (5.26) and (5.28) to estimate the seasonal state variable by inverting the frequency bounds. All corresponding figures are presented in the Appendix B of this chapter.

Figure 5.1 presents a summary of SIML smoothing for log-transformed data. This was done because the original series has a significant heteroscedastic seasonality. In Figs. 5.1, 5.3, and 5.4, "org" stands for the original series, "trend", "seasonal", and "noise" mean the estimated trend, seasonal, and noise components, while "adj" means the estimated seasonally adjusted series, i.e., the observed series minus the estimated seasonal component. "Z" denotes the transformed orthogonal series. In Figs. 5.2, 5.3, and 5.4, "reg" stands for the dummy variable.

The original time series exhibits the typical characteristics of major macroeconomic time series in Japan, i.e., it is a realization of non-stationary time series and displays a rather discernible trend, cycle, seasonal, and irregular components. We applied the SIML filtering with $m = 29$; the estimated trend-cycle component is indicated by the red curve. Given that $\lambda_m = 29/[2n] \sim 0.14$, which is approximately 1.8 years, that is, we have estimated the trend-cycle components over about 2 years cycle. This may be practically reasonable choice for the trend-cycle component

---

[3] We took the data from https://www.esri.cao.go.jp/jp/sna/menu.html (Economic and Social Research Institute (ESRI), Cabinet Office, Japan). They are original series in real terms and ESRI uses the X-12-ARIMA smoothing program for constructing seasonal adjusted official data. We use the consumption series of durable goods as a typical component of GDP.

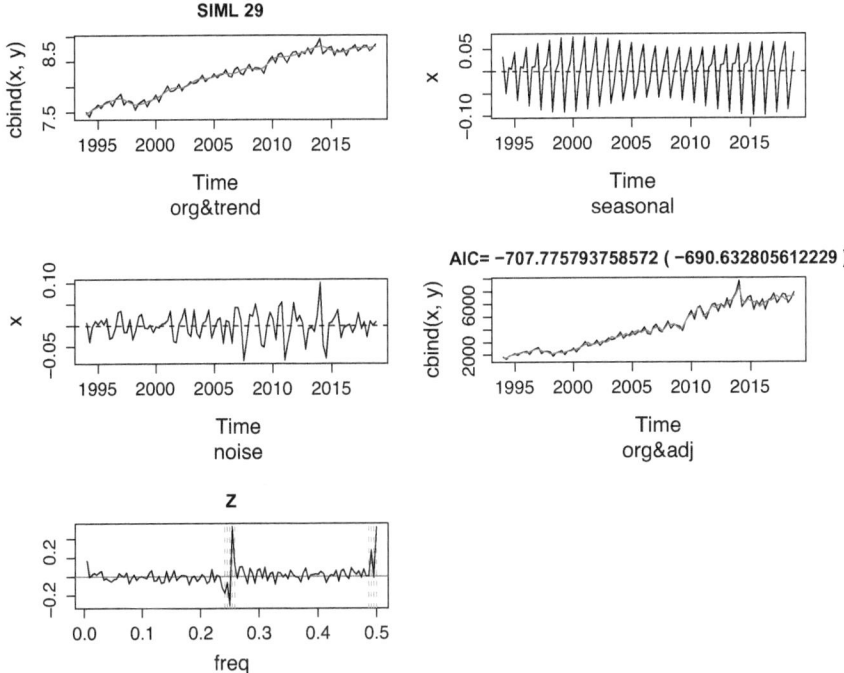

Fig. 5.1 Macro consumption (Data are the quarterly real consumption of durable goods (after log-transformation) between 1994Q1–2019Q4 published by the Economic Social Research Institute (ESRI), Cabinet Office, Japan.)

derived from the macroeconomic data. The noise component is constructed as the observed data minus the estimated trend-cycle and seasonal components. By using $\mathbf{Z}_n$-transformed data, we capture the significant effects at the seasonal frequencies. Because we use the quarterly data, we have sharp peaks and troughs at frequency 1/4 and 1/2. The estimated seasonal component moves regularly, which may change over time rather smoothly. We have found that the estimated seasonal component by using the X-13ARIMA-SEATS tend to exhibit more rigid seasonality. While the estimated seasonal component exhibits a regular seasonal pattern, the estimated trend-cycle and noise components indicate that there were some abrupt changes around the year of 2008–2009, 2011, and 2014, which are different from the typical noise component. (One way to deal with these effects is to use outlier detection and Reg-ARIMA model in X-13ARIMA-SEATS program.) It may be appropriate to consider the possibility that there are major breaks and institutional changes during the sample period.

First, there was a rapid downward effect attributed to the 2008 financial crisis occurred. This event may be considered appropriate for the application of the ramp-dummy at 2008Q3–2009Q1. Figures 5.2 and 5.3 provide a summary of the SIML smoothing and frequency regression results for the cases.

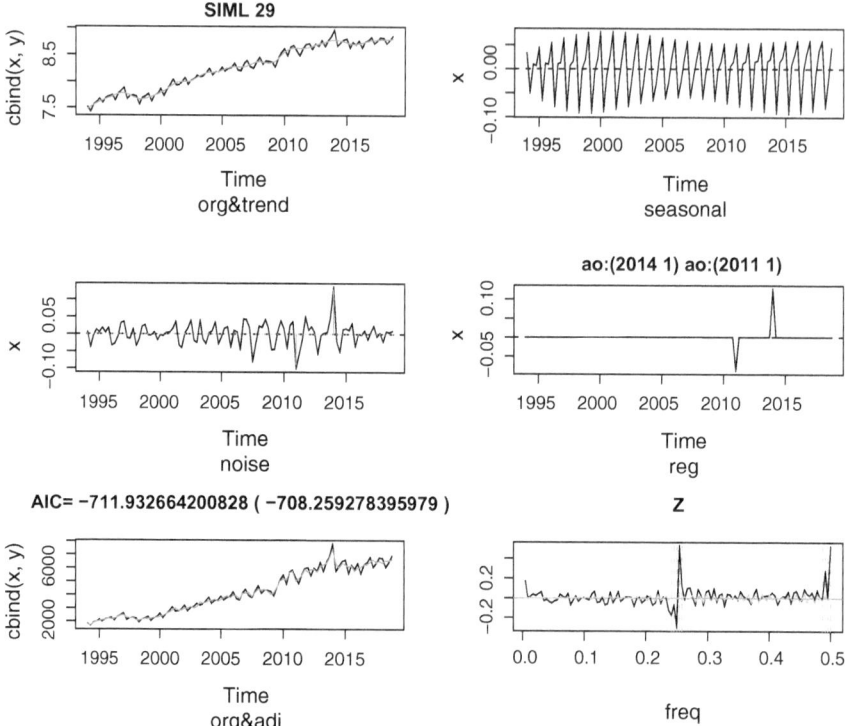

Fig. 5.2 Macro consumption (Data are the quarterly real consumption of durable goods (after log-transformation) between 1994Q1–2019Q4 published by the Economic Social Research Institute (ESRI), Cabinet Office, Japan.)

Secondly, we applied two AO-dummy variables at 2011Q1 and 2014Q1. (See Example 5.2 in Sect. 5.4.) In these periods, there were large effects caused by the 2011 earthquake in Japan and an increase of consumption tax introduced in April 2014. There was a temporary increase of durable consumption in the 2014Q1 period. Both events had significant effects on the macroeconomy and consumption in Japan.

Figure 5.4 represents a summary of SIML smoothing and frequency regression with three dummy variables considered simultaneously. (See Examples 5.2 and 5.4 in Sect. 5.4.) Based on the criteria of AIC, we selected the last case for the best modeling for the macro consumption of durable goods; these effects are captured by our method. By using the transformed data of (5.15) and (5.16) and the dummy variables, the AIC(w) was calculated based on the regression equation by

$$\text{AIC}(w) = n \log \hat{\sigma}_w^2 + 2r \tag{5.39}$$

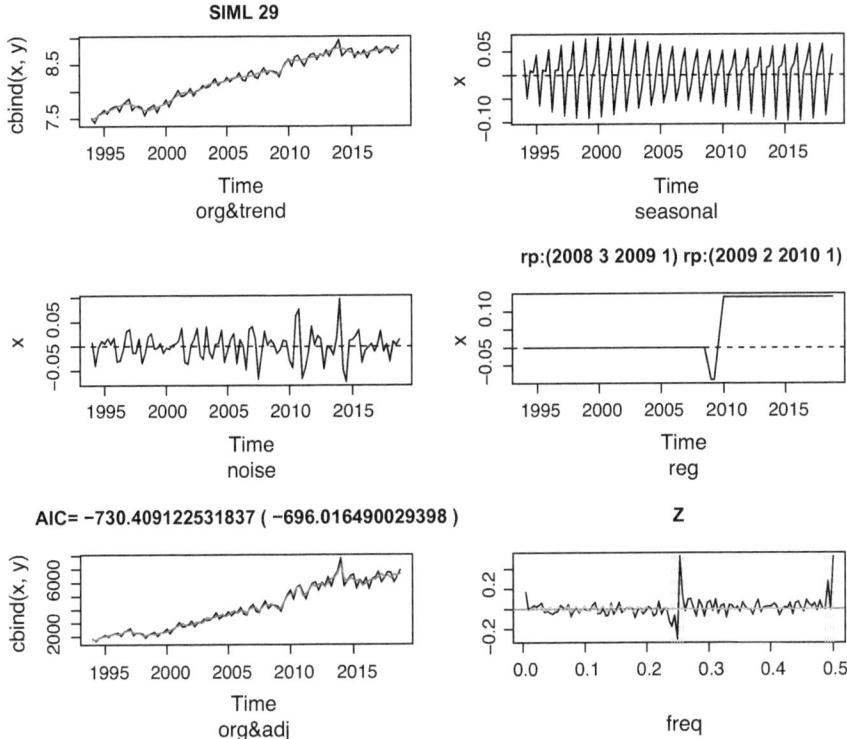

Fig. 5.3 Macro consumption (Data are the quarterly real consumption of durable goods (after log-transformation) between 1994Q1–2019Q4 published by the Economic Social Research Institute (ESRI), Cabinet Office, Japan.)

where we use $\hat{\sigma}_w^2$ calculated from the residuals of the dummy regression ((4.2) with $p = 1$) and r denotes the number of dummy variables.[4]

In our example, the main purpose of data analysis was to evaluate the appropriateness of the published data. This type of task was not easy because the published data used the X-12-ARIMA program of the U.S. Census Bureau and it is a complicated procedure in practice.[5] For each model, we have calculated two AICs: the first AIC in figures was calculated using all frequency data while the AIC in the parenthesis was calculated using all frequency data except data around the seasonal frequency.

---

[4] This AIC(w) is based on (5.36) and (5.37) with dummy variables, which can be implemented easily. However, we have taken the case as if a_{kn}^* were constant with respect to k because we use the procedure, that is free from the maximum likelihood (ML) estimation of unknown parameters needed. In this sense, our AIC(w) is an approximate one.

[5] The details of their estimation procedure from original series are explained at the web-cite of ESR (Economic and Social Research Institute, Cabinet office of Japan), https://www.esri.cao.go.jp/en/sna/sokuhou/sokuhou_top.html.

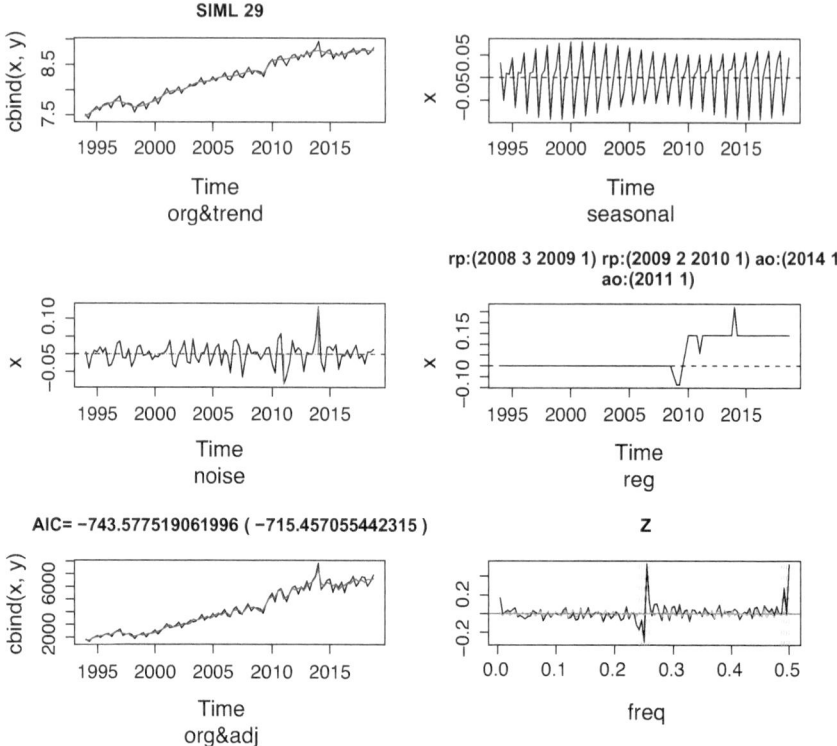

Fig. 5.4 Macro consumption (Data are the quarterly real consumption of durable goods (after log-transformation) between 1994Q1–2019Q4 published by the Economic Social Research Institute (ESRI), Cabinet Office, Japan.)

By using the model selection criteria for minimizing these AICs, we find that SIML smoothing with three dummy variables (i.e., two AOs and a double ramp) is the best model. We have reasonable result on the decomposition of original time series into trend-cycle, seasonal, and noise components. In the selected model, the trend-cycle component includes one structural change and the noise component includes two outliers.

This empirical analysis demonstrates the necessity of considering the role of incorporating the effects of the change-point problem and abrupt changes in the seasonal adjustment procedure. In this regard, we have demonstrated our methodology based on the frequency regression and smoothing.

5.7 Some Remarks

In a considerable number of original macroeconomic time series, it is common to observe non-stationary trend, cycles, seasonal, and measurement error components simultaneously. In addition to these components, we sometimes observe abrupt changes, trading-day effects, and other irregular components. It is therefore challenging to remove the seasonal component from the original time series in the seasonal adjustment and construct a macro-index, incorporating several non-stationary time series.

This chapter introduces a new approach for dealing with non-stationary time series using the frequency regression based on the SIML modeling in a systematic manner. We use the SIML method because we can separate the likelihood information of time series data into different frequency parts of their components. Our method sheds a new light on some practical approach to handle economic time series, which have been practically used in official seasonal adjustments without formal justification. There will be many empirical examples.

There are further problems to be studied. The present SIML method is based on the additive time series decomposition in (5.1)–(5.2). There can be more complicated decomposition models including trends, cycles, and seasonal components in different forms. Then, we need to investigate the relationships among trends, cycles, seasonal, and irregular noise components in non-stationary and stationary time series both in the time and frequency domains. It appears that some extensions of Theorems 5.1 and 5.2 in this chapter can be developed.

Appendix: Mathematical Derivations

We present the derivations of Theorems 5.1 and 5.2 as an application of Theorem 5.3 below. Theorem 5.3 gives the asymptotic properties of the estimation of long-run variance-covariance matrix for the non-stationary errors-in-variables models. We first provide the intuition for our result and then give some details of its derivation. This Appendix reports some improvements of the results reported in Chap. 3 and Kunitomo and Sato (2021).

A-I A Heuristic Derivation of Propositions in Chap. 3 and Theorems in This Chapter

We consider the non-stationary process $\mathbf{y}_i$ $(i = 1, \ldots, n)$ defined by (5.1)–(5.2) given the initial conditions $\mathbf{y}_i, \mathbf{x}_i$ $(i \leq 0)$. (Without loss of generality, we often ignore the effects of initial conditions whenever they are asymptotically negligible.)

Let $\theta_{jk} = \frac{2\pi}{2n+1}(j - \frac{1}{2})(k - \frac{1}{2})$, $p_{jk}^{(n)} = \frac{1}{\sqrt{2n+1}}(e^{i\theta_{jk}} + e^{-i\theta_{jk}})$ and for $\mathbf{Y}_n = (\mathbf{y}_i')$ we write $\mathbf{z}_k$ $(k = 1, \ldots, n)$ as

$$\mathbf{z}_k(\lambda_k^{(n)}) = \sum_{j=1}^{n} p_{jk}^{(n)} \mathbf{r}_j, \quad \mathbf{r}_j = \mathbf{y}_j - \mathbf{y}_{j-1}, \tag{5.40}$$

which is a (real-valued) Fourier transform and the initial conditions are fixed.

Then, we find that $\mathbf{z}_k(\lambda_k^{(n)}) = (\mathbf{z}_{in}(\lambda_k^{(n)}))$ $(k = 1, \ldots, n)$ are the (real-valued) Fourier transform of data at the frequency $\lambda_k^{(n)}$ $(= (k - 1/2)/(2n + 1))$, which is a (real-part of) estimate of the orthogonal incremental vector process $(p \times 1)$ $\mathbf{z}(\lambda) = (\mathbf{z}_i(\lambda))$ $(0 \le \lambda \le 1/2)$ and $\mathbf{z}(\lambda)$ is a continuous process in the frequency domain.

By evaluating

$$\mathbf{E}\left[\mathbf{z}_k(\lambda_k^{(n)})\mathbf{z}_{k'}(\lambda_{k'}^{(n)})'\right] = \left[\frac{1}{2n+1}\right] \sum_{j,j'=1}^{n} (e^{i\theta_{jk}} + e^{-i\theta_{jk}})(e^{i\theta_{j'k'}} + e^{-i\theta_{j'k'}})\mathscr{E}[\mathbf{r}_j\mathbf{r}_{j'}'],$$

we find that the effects of each term with $k \neq k'$ are asymptotically negligible, and the dominant sum with $k = k'$ is asymptotically equivalent to

$$\left[\frac{n}{2n+1}\right] \sum_{h=-(n-1)}^{n-1} \left[\cos 2\pi \frac{k - 1/2}{2n+1} h\right] [\mathbf{\Gamma}(h) + \mathbf{\Gamma}(-h)],$$

where we use the notation $\mathbf{\Gamma}(h) = \mathbf{E}(\mathbf{r}_j\mathbf{r}_{j-h}')$ (we have ignored the constant means with $\mathbf{E}[\mathbf{r}_j] = \mathbf{0}$ as in the proof of Proposition 3.1).

Then, by using the conditions in (5.1)–(5.2), it converges to

$$\mathbf{f}_{SR}(\lambda) = \sum_{h=-\infty}^{\infty} \cos(2\pi h\lambda)\mathbf{\Gamma}(h), \quad 0 \le \lambda \le \frac{1}{2}, \tag{5.41}$$

where the symmetrized spectral density matrix for $\mathbf{r}_j$ is given by $\mathbf{f}_{SR}(\lambda) = (1/2)[\mathbf{f}_{\Delta y}(\lambda) + \bar{\mathbf{f}}_{\Delta y}(\lambda)]$.

Under the assumption of stationarity of $\mathbf{r}_j$ it has been known that $\mathbf{z}_k(\lambda_k^{(n)})$ are asymptotically uncorrelated random variables. (See Proposition 3.1 of Chap. 3 and Chaps. 8–9 of Anderson (1971), Brockwell and Davis (1990), for instance.) By using straightforward (but lengthy) evaluations, we find that for $k \neq k'$

$$\mathbf{E}[\mathbf{z}_{ik}(\lambda_k^{(n)})\mathbf{z}_{jk}(\lambda_k^{(n)})\mathbf{z}_{hk}(\lambda_{k'}^{(n)})\mathbf{z}_{lk}(\lambda_{k'}^{(n)})] = \sigma_{ij}(\lambda_k^{(n)})\sigma_{hl}(\lambda_{k'}^{(n)}) + o(1) \tag{5.42}$$

and for $k = k'$

$$\mathbf{E}[\mathbf{z}_{ik}(\lambda_k^{(n)})\mathbf{z}_{jk}(\lambda_k^{(n)})\mathbf{z}_{hk}(\lambda_k^{(n)})\mathbf{z}_{lk}(\lambda_k^{(n)})] \tag{5.43}$$
$$= \sigma_{ij}(\lambda_k^{(n)})\sigma_{hl}(\lambda_k^{(n)}) + \sigma_{ih}(\lambda_k^{(n)})\sigma_{jl}(\lambda_k^{(n)}) + \sigma_{il}(\lambda_k^{(n)})\sigma_{jh}(\lambda_k^{(n)}) + o(1),$$

where $\mathbf{\Gamma}(h) = (\Gamma_{ij}(h))$ and

$$\sigma_{ij}(\lambda_k^{(n)}) = \sum_{h=-(n-1)}^{n-1} \left[\cos 2\pi \lambda_k^{(n)} h\right] \Gamma_{ij}(h). \tag{5.44}$$

From (5.42)–(5.43), we notice that $\mathbf{z}_k(\lambda_k^{(n)})$ are (as if) normally distributed random variables with (5.41).

As $n \to \infty$ and $m/n \to 0$, we have $\lambda_k^{(n)} \to 0$ for $1 \le k \le m$. We write for $k = 1, \ldots, m$ and as $m/n \to 0$,

$$\lim_{n \to \infty} \sigma_{ij}(\lambda_k^{(n)}) = \sigma_{ij}^{(x)} \ (i, j = 1, \ldots, p) \tag{5.45}$$

and $\mathbf{\Sigma}_x = (\sigma_{ij}^{(x)})$. Then in this situation

$$\mathbf{Var}\left[\frac{1}{\sqrt{m}} \sum_{k=1}^m \mathbf{z}_{ik}(\lambda_k^{(n)})\mathbf{z}_{jk}(\lambda_k^{(n)})\right] \longrightarrow \sigma_{ii}^{(x)}\sigma_{jj}^{(x)} + \sigma_{ij}^{(x)2}. \tag{5.46}$$

We construct a sequence of random variables, which are approximately uncorrelated and for $i, j = 1, \ldots, p$

$$s_{ij}(t) = \mathbf{z}_{ik}(\lambda_t^{(n)})\mathbf{z}_{jk}(\lambda_t^{(n)}) - \mathscr{E}[\mathbf{z}_{ik}(\lambda_t^{(n)})\mathbf{z}_{jk}(\lambda_t^{(n)})]$$

and

$$M_{ij}(n, k) = \sum_{t=1}^k s_{ij}(t).$$

Then, heuristically, we can apply the central limit theorem (CLT) for the multivariate Gaussian stationary process to obtain the asymptotic normality of the normalized quadratic quantities. However, to show this argument in a rigorous way, we need further developments.

A-II Proof of Main Results: We first prepare a general result on the consistency and asymptotic normality of the SIML estimation in non-stationary time series; the result may have some new aspect.

Theorem 5.3 *Assume that the fourth-order moments of each element of* $\mathbf{v}_i^{(x)}$ *and* $\mathbf{v}_i$ *in (5.1)–(5.2) are bounded. Let*

$$\hat{\mathbf{\Sigma}}_x\left(= (\hat{\sigma}_{gh}^{(x)})\right) = \frac{1}{m}\mathbf{Z}_m^{*'}\mathbf{Z}_m^*, \tag{5.47}$$

which is $\mathbf{G}_m^*$ *in (5.5). Then*

(i) For $m_n = n^\alpha$ ($[m_n] = m$) and $0 < \alpha < 1$, as $n \longrightarrow \infty$

$$\hat{\boldsymbol{\Sigma}}_x - \boldsymbol{\Sigma}_x \xrightarrow{p} \mathbf{O}. \tag{5.48}$$

(ii) We set $\boldsymbol{\Sigma}_x = (\sigma_{gh}^{(x)})$. For $m_n = [n^\alpha]$ and $0 < \alpha < 0.8$, as $n \longrightarrow \infty$

$$\sqrt{m_n}\left[\hat{\sigma}_{gh}^{(x)} - \sigma_{gh}^{(x)}\right] \xrightarrow{\mathscr{L}} N\left(0, \sigma_{gg}^{(x)}\sigma_{hh}^{(x)} + \left[\sigma_{gh}^{(x)}\right]^2\right). \tag{5.49}$$

The covariance of the limiting distributions of $\sqrt{m_n}[\hat{\sigma}_{gh}^{(x)} - \sigma_{gh}^{(x)}]$ and $\sqrt{m_n}[\hat{\sigma}_{kl}^{(x)} - \sigma_{kl}^{(x)}]$ is given by $\sigma_{gk}^{(x)}\sigma_{hl}^{(x)} + \sigma_{gl}^{(x)}\sigma_{hk}^{(x)}$ ($g, h, k, l = 1, \ldots, p$).

Proof of Theorem 5.3 The proof consists of several steps.

(Step 1): Let $\mathbf{z}_k^{(x)} = (z_{kj}^{(x)})$ and $Z_k^{(v)} = (z_{kj}^{(v)})$ ($k = 1, \ldots, n$) be the k-th row vector elements of $n \times p$ matrices

$$\mathbf{Z}_n^{(x)} = \mathbf{K}_n^*(\mathbf{X}_n - \bar{\mathbf{X}}_0), \quad \mathbf{Z}_n^{(v)} = \mathbf{K}_n^*\mathbf{V}_n, \quad \mathbf{K}_n^* = \mathbf{P}_n\mathbf{C}_n^{-1}, \tag{5.50}$$

respectively, where we denote $\mathbf{X}_n = (\mathbf{x}_k') = (x_{kg}), \mathbf{V}_n = (\mathbf{v}_k') = (v_{kg}), \mathbf{Z}_n = (\mathbf{z}_k') (= (z_{kg}))$ as $n \times p$ matrices with $z_{kg} = z_{kg}^{(x)} + z_{kg}^{(v)}$ Here, we set the initial conditions $\bar{\mathbf{X}}_0 = \bar{\mathbf{Y}}_0$ and we find that the effects of initial condition are stochastically negligible in the frequency regression. We write $z_{kg}, z_{kg}^{(x)}, z_{kg}^{(v)}$ as the g-th component of $\mathbf{z}_k, \mathbf{z}_k^{(x)}$, and $\mathbf{z}_k^{(v)}$ ($k = 1, \ldots, n; g = 1, \ldots, p$).
(Note that the above notations of subscripts are slightly different from $\mathbf{y}_i = (y_{ji})$, ($j = 1 \ldots, p; i = 1, \ldots, n$ in this section.)
    By decomposing $\hat{\boldsymbol{\Sigma}}_x - \boldsymbol{\Sigma}_x (= (\hat{\sigma}_{gh}^{(x)} - \sigma_{gh}^{(x)})$ for $g, h = 1, \ldots, p$) into the effects of $\mathbf{z}_k^{(x)}$ and $\mathbf{z}_k^{(v)}$, we rewrite

$$\sqrt{m_n}\left[\frac{1}{m}\sum_{k=1}^{m}\mathbf{z}_k\mathbf{z}_k' - \boldsymbol{\Sigma}_x\right] \tag{5.51}$$

$$= \sqrt{m_n}\left[\frac{1}{m}\sum_{k=1}^{m}\mathbf{z}_k^{(x)}\mathbf{z}_k^{(x)'} - \boldsymbol{\Sigma}_x\right] + \frac{\sqrt{m_n}}{m}\sum_{k=1}^{m}\mathbf{E}[\mathbf{z}_k^{(v)}\mathbf{z}_k^{(v)'}]$$

$$+ \frac{\sqrt{m_n}}{m}\sum_{k=1}^{m}\left[\mathbf{z}_k^{(v)}\mathbf{z}_k^{(v)'} - \mathbf{E}[\mathbf{z}_k^{(v)}\mathbf{z}_k^{(v)'}]\right] + \frac{\sqrt{m_n}}{m}\sum_{k=1}^{m}\left[\mathbf{z}_k^{(x)}\mathbf{z}_k^{(v)'} + \mathbf{z}_k^{(v)}\mathbf{z}_k^{(x)'}\right].$$

Then we shall show that three terms except the first term of (5.51) in the right-hand side are $o_p(1)$ under the condition of $0 < \alpha < 0.8$. To show (ii) of Theorem 5.3, we need to show that the dominant term in (5.51) is

$$\sqrt{m_n}\left[\frac{1}{m}\sum_{k=1}^{m}\mathbf{z}_k^{(x)}\mathbf{z}_k^{(x)'} - \boldsymbol{\Sigma}_x\right], \tag{5.52}$$

and it is asymptotically normal as $m_n \to \infty$ $(n \to \infty)$. By denoting $\boldsymbol{\Gamma}_x(h) = \mathcal{E}[\Delta\mathbf{x}_t\Delta\mathbf{x}'_{t-h}]$, we express

$$\boldsymbol{\Sigma}_x = \mathbf{f}_{\Delta x}(0) = \sum_{h=-\infty}^{+\infty}\boldsymbol{\Gamma}_x(h). \tag{5.53}$$

From $\theta_{jk} = [2\pi/(2n+1)](j-\frac{1}{2})(k-\frac{1}{2})$, we set $c_{ij} = (2/m)\sum_{k=1}^{m}\cos(\theta_{ik})\cos(\theta_{jk})$ $(i,j=1,\ldots,n)$. Then for any (non-zero $p \times 1$) constant vector $\boldsymbol{a}$, we can evaluate

$$\mathbf{E}\left[\frac{1}{m}\sum_{k=1}^{m}(\boldsymbol{a}'\mathbf{z}_k^{(x)})^2 - \boldsymbol{a}'\boldsymbol{\Sigma}_x\boldsymbol{a}\right]^2 = \left(\frac{2}{2n+1}\right)^2\mathcal{E}\left[\sum_{j,j'=1}^{n}c_{jj'}\boldsymbol{a}'(\mathbf{v}_j^{(x)}\mathbf{v}_{j'}^{(x)'} - \mathbf{E}(\mathbf{v}_j^{(x)}\mathbf{v}_{j'}^{(x)'})\boldsymbol{a}\right]^2$$

$$\leq K_1\left[\frac{2}{2n+1}\right]^2\sum_{j,j'=1}^{n}c_{j,j'}^2,$$

where K_1 is a positive constant and we have used the boundedness of fourth moments. Since $m\sum_{j,j'=1}^{n}c_{j,j'}^2 = (n+1/2)^2$ (Lemma 5.2 of Kunitomo et al. (2018)), we can show that the first term is $O_p(1)$ as $m \to \infty$, $m/n \to 0$.

(**Step 2**): Let $\mathbf{b}_k = (b_{kj}) = \boldsymbol{\alpha}'_k\mathbf{P}_n\mathbf{C}_n^{-1} = (b_{kj})$ and $\boldsymbol{\alpha}_k^{(n)'} = (0,\ldots,1,0,\ldots)$ be an $n \times 1$ vector. We write $z_{kg}^{(v)} = \sum_{j=1}^{n}b_{kj}v_{jg}$ for the seasonal and noise part and use the relation

$$(\mathbf{P}_n\mathbf{C}_n^{-1}\mathbf{C}_n'^{-1}\mathbf{P}_n')_{k,k'} = \delta(k,k')4\sin^2\left[\frac{\pi}{2n+1}(k-\frac{1}{2})\right] = \delta(k,k')a_{kn}^*. \tag{5.54}$$

Then under the conditions that $\|\mathbf{C}_j^{(v)}\| = O(\rho^j)$ $(0 \leq \rho < 1)$, we can find K_2 (a positive constant) such that

$$\mathbf{E}\left[(z_{kg}^{(v)})\right]^2 = \mathcal{E}[\sum_{i=1}^{n}b_{ki}v_{ig}\sum_{j=1}^{n}b_{kj}v_{jg}] \leq K_2 \times a_{kn}^*. \tag{5.55}$$

It is because $\mathbf{E}[(z_{kg}^{(v)})]^2 = \sum_{i,j=1}^{n}b_{ki}b_{kj}\sigma_{gg}^{(v)}(i-j)$, where $\sigma_{gg}^{(v)}(i-j)$ is the $(i-j)$-th autocovariance of v_{ig} and v_{jg}. (We denote $b_{ki} = 0$ for $i < 0$ and $i > n$.) Then

$$\mathbf{E}[(z_{kg}^{(v)})]^2 = \sum_{l=-(n-1)}^{n-1}\left[\sum_{j=1}^{n}b_{kj}b_{k,j+l}\sigma_{gg}^{(v)}(l)\right] \leq \left[\sum_{j=1}^{n}b_{kj}^2\right]\sum_{l=-\infty}^{\infty}|\sigma_{gg}^{(v)}(l)|.$$

Because $\|\mathbf{C}_j^{(v)}\| = O(\rho^j)$, $\sum_{l=-\infty}^{\infty} |\sigma_{gg}^{(v)}(l)|$ is bounded. Also it is straightforward to find that

$$\frac{1}{m} \sum_{k=1}^{m} a_{kn}^* = \frac{1}{m} 2 \sum_{k=1}^{m} \left[1 - \cos\left(\pi \frac{2k-1}{2n+1} \right) \right] = O\left(\frac{m^2}{n^2} \right), \tag{5.56}$$

by using the relation

$$\sum_{k=1}^{m} 2 \cos\left(\pi \frac{2k-1}{2n+1} \right) = \sum_{k=1}^{m} \left[e^{i \frac{2\pi}{2n+1}(k-\frac{1}{2})} + e^{-i \frac{2\pi}{2n+1}(k-\frac{1}{2})} \right] = \frac{\sin\left(\frac{2\pi}{2n+1} m \right)}{\sin\left(\frac{\pi}{2n+1} \right)}.$$

Then the second term of (5.51) becomes

$$\frac{1}{m} \sum_{k=1}^{m} \mathbf{E}[z_{kg}^{(v)}]^2 \leq K_3 \frac{1}{m} \sum_{k=1}^{m} a_{kn}^* = O\left(\frac{m^2}{n^2} \right), \tag{5.57}$$

which is $o(1)$ if we set α such that $0 < \alpha < 1$ and K_3 is a positive constant. Hence, we find that

$$\frac{1}{\sqrt{m}} \sum_{k=1}^{m} \mathbf{E}[z_{kg}^{(v)}]^2 \leq K_3 \frac{1}{\sqrt{m}} \sum_{k=1}^{m} a_{kn}^* = O_p\left(\frac{m^{5/2}}{n^2} \right).$$

Then it is $o_p(1)$ if $0 < 5/2\alpha < 2$, that is, $0 < \alpha < 0.8$. (This condition is needed for the asymptotic normality.)

For the fourth term,

$$\mathbf{E}\left[\frac{1}{m} \sum_{j=1}^{m} z_{kg}^{(x)} z_{kg}^{(v)} \right]^2 = \frac{1}{m^2} \sum_{k,k'=1}^{m} \mathbf{E}\left[z_{kg}^{(x)} z_{k',g}^{(x)} z_{kg}^{(v)} z_{k',g}^{(v)} \right] = O\left(\frac{m}{n^2} \right).$$

In the above evaluation we have used the evaluation that if we set $s_{jk} = \cos\theta_{jk}$ $(j, k = 1, 2, \ldots, n)$, then we have the relation $\sum_{j=1}^{n} s_{jk}^2 = \frac{n}{2} + \frac{1}{4}$ by a direct calculation (see Lemma 5.1 of Kunitomo et al. (2018)) that

$$\left| \sum_{j=1}^{n} s_{jk} s_{j,k'} \right| \leq \left[\sum_{j=1}^{n} s_{jk}^2 \right] = \frac{n}{2} + \frac{1}{4} \quad \text{for any } k \geq 1.$$

Finally, for the third term, we need to consider the variance of

$$(z_{kg}^{(v)})^2 - \mathscr{E}[(z_{kg}^{(v)})^2] = \sum_{j,j'=1}^{n} b_{kj} b_{k,j'} \left[v_{jg} v_{j'g} - \mathscr{E}[v_{jg} v_{j',g}] \right].$$

Then by using the assumptions, after some evaluations, we find a positive constant K_4 such that

$$\mathbf{E}\left[\frac{1}{m_n}\sum_{k=1}^{m}((z_{kg}^{(v)})^2 - \mathscr{E}[(z_{kg}^{(v)})^2])\right]^2$$

$$= \frac{1}{m^2}\sum_{k_1,k_2=1}^{m}\mathbf{E}\left[\sum_{j_1,j_2,j_3,j_4=1}^{n} b_{k_1,j_1}b_{k_1,j_2}\left(v_{j_1,g}v_{j_2,g} - \mathscr{E}(v_{j_1,g}v_{j_2,g})\right)\right.$$

$$\left. \times b_{k_2,j_3}b_{k_2,j_4}\left(v_{j_3,g}v_{j_4,g} - \mathscr{E}(v_{j_3,g}v_{j_4,g})\right)\right]$$

$$\leq K_4 \frac{1}{m^2}\left[\sum_{k=1}^{m}a_{kn}^*\right]^2 = O\left(\frac{1}{m^2}\times\left(\frac{m^3}{n^2}\right)^2\right), \tag{5.58}$$

which is $O(m^4/n^4)$ by straightforward calculations. It is straightforward, but tedious to show the evaluation and then here we just give an illustration of our derivations when $p = 1$ and we rewrite $v_i = \sum_{j=0}^{\infty} c_j^{(v)}e_{i-j}^{(v)}$ $(c_j^{(v)} = \mathbf{C}_j^{(v)})$ and we evaluate

$$\sum_{k_1,k_2=1}^{m}\sum_{j_1,j_2,j_3,j_4} b_{k_1,j_1}b_{k_1,j_2}b_{k_2,j_3}b_{k_2,j_4}\times \mathbf{E}\{[v_{j_1}v_{j_2} - \mathscr{E}(v_{j_1}v_{j_2})][v_{j_3}v_{j_4} - \mathscr{E}(v_{j_3}v_{j_4})]\}$$

$$= \sum_{k_1,k_2=1}^{m}\sum_{j_1,j_2,j_3,j_4} b_{k_1,j_1}b_{k_1,j_2}b_{k_2,j_3}b_{k_2,j_4}\sum_{l_1,l_2,l_3,l_4=0}^{\infty} c_{l_1}^{(v)}c_{l_2}^{(v)}c_{l_3}^{(v)}c_{l_4}^{(v)}$$

$$\times \mathbf{E}\{[e_{j_1-l_1}e_{j_2-l_2} - \mathscr{E}(e_{j_1-l_1}e_{j_2-l_2})][e_{j_3-l_3}e_{j_4-l_4} - \mathscr{E}(e_{j_3-l_3}e_{j_4-l_4})]\}.$$

We need to evaluate the corresponding terms for four cases when (i) $j_1 - l_1 = j_2 - l_2 = j_3 - l_3 = j_4 - l_4$, (ii) $j_1 - l_1 = j_2 - l_2 \neq j_3 - l_3 = j_4 - l_4$, (iii) $j_1 - l_1 = j_3 - l_3 \neq j_2 - l_2 = j_4 - l_4$, (iv) $j_1 - l_1 = j_4 - l_4 \neq j_2 - l_2 = j_4 - l_4$. By using the condition in the general case that $\|\mathbf{C}_j^{(v)}\| = O(\rho^j)$ $(j \geq 0, 0 \leq \rho < 1)$, we have $\sum_{j=0}^{\infty}|c_j^{(v)}| < \infty$ in this special case. We also utilize the relation such as $\sum_{j=1}^{n} b_{kj}b_{k'j} = \delta(k,k')a_{kn}^*$ in the general case and we have the notation that $b_{k,j} = 0$ for $k = 1,\ldots,m$, $j < 0$, $j > n$, and $c_j = 0$ $(j < 0)$.

Then in each (i)–(iv) case, we can take a positive constant K_5 such that (5.58) is less than

$$K_5 \sum_{k_1,k_2=1}^{m}\left[\sum_{j_1=1}^{n}b_{k_1,j_1}^2\right]^{1/2}\left[\sum_{j_2=1}^{n}b_{k_1,j_2}^2\right]^{1/2}\left[\sum_{j_3=1}^{n}b_{k_2,j_3}^2\right]^{1/2}\left[\sum_{j_4=1}^{n}b_{k_2,j_4}^2\right]^{1/2}.$$

The above evaluation method works in the general case with a complication of notations. Therefore, by using (5.58), the third term of (5.51) is negligible if we set α such that $0 \leq \alpha < 1$. (The derivations are similar to the ones in Kunitomo and Sato (2021).)

For the consistency of the SIML estimation, instead of (5.51), we use the representation

$$\frac{1}{m}\sum_{k=1}^{m}\mathbf{z}_k\mathbf{z}_k' - \mathbf{\Sigma}_x = \left[\frac{1}{m}\sum_{k=1}^{m}\mathbf{z}_k^{(x)}\mathbf{z}_k^{(x)'} - \mathbf{\Sigma}_x\right] + \frac{1}{m}\sum_{k=1}^{m}\mathbf{E}[\mathbf{z}_k^{(v)}\mathbf{z}_k^{(v)'}]$$
$$+\frac{1}{m}\sum_{k=1}^{m}\left[\mathbf{z}_k^{(v)}\mathbf{z}_k^{(v)'} - \mathbf{E}[\mathbf{z}_k^{(v)}\mathbf{z}_k^{(v)'}]\right] + \frac{1}{m}\sum_{k=1}^{m}\left[\mathbf{z}_k^{(x)}\mathbf{z}_k^{(v)'} + \mathbf{z}_k^{(v)}\mathbf{z}_k^{(x)'}\right].$$

Then, the second, third, and fourth terms converge to 0 if we have the condition $0 < \alpha < 1$. (For the asymptotic normality, we need the condition $0 < \alpha < 0.8$.)

(**Step 3**): We need to evaluate the limiting distribution of the first term of (5.51). Instead of (5.52), however, we consider the asymptotic distribution of

$$s_{ij}^{(m)*} = \frac{1}{\sqrt{m}}[g_{ij}^{(m*)} - \mathbf{E}(g_{ij}^{(m*)})] \tag{5.59}$$

and

$$g_{ij}^{(m*)} = \left(\frac{1}{m}\sum_{k=1}^{m}\mathbf{z}_k^{(x)}\mathbf{z}_k^{(x)'}\right)_{ij} \quad (i, j = 1, \ldots, p). \tag{5.60}$$

By using $\mathbf{P}_n = (p_{jk}^{(n)})$, we decompose

$$s_{ij}^{(m)*} = \frac{1}{\sqrt{m}}\sum_{k=1}^{m}\left[\sum_{s=t=1}^{n} p_{ks}^{(n)2}(v_{is}^{(x)}v_{js}^{(x)} - \mathbf{E}(v_{is}^{(x)}v_{js}^{(x)}))\right] \tag{5.61}$$
$$+\frac{1}{\sqrt{m}}\sum_{k=1}^{m}\left[\sum_{s\neq t=1}^{n} p_{ks}^{(n)}p_{kt}^{(n)}(v_{is}^{(x)}v_{jt}^{(x)} - \mathbf{E}(v_{is}^{(x)}v_{jt}^{(x)}))\right]$$

and $\mathbf{v}_t^{(x)} (= (v_{it}^{(x)})) = \Delta\mathbf{x}_t = \sum_{s=0}^{\infty}\mathbf{C}_s^{(x)}\mathbf{e}_{t-s}^{(x)}$, where $\mathbf{C}_s^{(x)} (= (C_{is}))$ are $p \times p$ matrices with $C_{is} = O(\rho^{|h|})$ $(0 \leq \rho < 1)$, and $\mathbf{e}_s^{(x)}$ are a sequence of mutually independent random vectors with $\mathbf{E}[\mathbf{e}_s^{(x)}] = \mathbf{0}$, $\mathbf{E}[\mathbf{e}_s^{(x)}\mathbf{e}_s^{(x)'}] = \mathbf{\Sigma}_v^{(x)} (> 0)$.

When we have the condition $m_n/n \to 0$ as $n \to \infty$, we have $\frac{1}{\sqrt{m_n}}[\sigma_{ij}^{(x)} - \mathbf{E}(g_{ij}^{(m*)})] = o(1)$. The evaluation of the limiting distribution of (5.61) is considerably simpler than that for (5.52).

We use the representations $p_{ks}^{(n)2} = [4/(2n+1)][\cos\theta_{ks}]^2$, $\sum_{k=1}^{m} p_{ks}^{(n)2} = [2m/(2n+1)]c_{ss}$, and $\sum_{s=1}^{n} c_{ss}^2 = O(n)$, $c_{st} = [2/m]\sum_{k=1}^{m}\cos\theta_{sk}\cos\theta_{tk}$, and $\theta_{jk} = \frac{2\pi}{2n+1}(j - \frac{1}{2})(k - \frac{1}{2})$. Because c_{ss} in bounded, and $\mathbf{v}_s^{(x)}$ has a MA representation with conditions on its coefficients in (2.11), it is possible to evaluate the variances of $[2\sqrt{m}/(2n+1)]\sum_{s=1}^{n} c_{ss}[v_{is}^{(x)}v_{jt}^{(v)} - \mathbf{E}(v_{is}^{(v)}v_{jt}^{(x)})]$, which converge to zeros in probability when $m_n/n \to 0$ as $n \to \infty$. Hence, the first term of (5.61) is asymptotically negligible because of the condition $m_n/n \to 0$ as $n \to \infty$.

Then, by using symmetry, we only need to show the asymptotic normality of the leading term of (5.61), which is

$$s_{ij}^{(m)**} = \frac{2\sqrt{m}}{2n+1} \sum_{s \neq t=1}^{n} c_{st}[v_{is}^{(x)} v_{jt}^{(v)} - \mathbf{E}(v_{is}^{(v)} v_{jt}^{(x)})]. \tag{5.62}$$

Under the stationarity condition of $\mathbf{v}_s^{(x)}$, the difference between (5.62) and the second term of (5.51) is asymptotically negligible. Also under the stationarity and the conditions on coefficients in (2.11) and (2.12), it has been known in time series analysis that the effects of initial conditions on $\mathbf{v}_s^{(x)}$ ($s \leq 0$) are asymptotically negligible. (We omit the detail of this argument because it may be straightforward.)

(**Step 4**): Our proof of the asymptotic normality requires a further derivation, which is a modification of the method for the spectral density estimation used in the proof of Theorem 9.4.1 in Anderson (1971). Because some of our arguments are similar, we only repeat the essential arguments and some differences. We provide the proof for the case when $p = 1$ and use the notation $\mathbf{C}_s^{(x)} = c_s$ ($s = 0, 1, \ldots$), $\mathbf{v}_s^{(x)} = v_s^{(x)}$, $\mathbf{e}_s^{(x)} = e_s$, and $s^{(m)**} = s_{ij}^{(m)**}$ ($i = j = 1$) because the proof of the general case when $p \geq 1$ can be obtained by using the standard device of $v_j^* = \mathbf{a}' \mathbf{v}_j^{(x)}$ ($j = 1, \ldots, n$) with an arbitrary ($p \times 1$ non-zero constant) vector $\mathbf{a}$.

We take $K_n = [n/m_n]$ be a sequence of positive integers and $K_n \to \infty$ ($n \to \infty$). Then, given s, $c_{st} \to 0$ for $t - s > K_n$ as $m_n, n \to \infty$ and $m_n/n \to 0$. Then, by taking $t = s + k$ ($k = 1, \ldots, n - s$) we rewrite

$$s^{(m)**} = \frac{4\sqrt{m}}{2n+1} \sum_{t>s=1}^{n} c_{st}[v_s^{(x)} v_t^{(x)} - \mathbf{E}(v_s^{(x)} v_t^{(x)})], \tag{5.63}$$

$$= \frac{4\sqrt{m}}{2n+1} \sum_{l,l'=0}^{\infty} c_l c_{l'} \sum_{s=1}^{n} \sum_{k=1}^{n-s} c_{s,s+k}[e_{s-l} e_{s+k-l'} - \mathbf{E}(e_{s-l} e_{s+k-l'})].$$

We truncate the sum $\sum_{l,l'=1}^{\infty}[\cdot]$ by a sub-sequence r_n ($r_n \to \infty$ as $n \to \infty$) and decompose the sum as $(\sum_{l=1}^{r_n} + \sum_{l=r_n+1}^{\infty})(\sum_{l'=1}^{r_n} + \sum_{l'=r_n+1}^{\infty})[\cdot]$. We can take a sequence of sums $\sum_{l,l'=1}^{r_n}[\cdot]$ such that $r_n \to \infty$ and $\sum_{l=r_n+1}^{\infty} |\gamma_l| \to 0$. Then we approximate the infinite sum by a finite sum because the remaining terms are of smaller order asymptotically. The main term is

$$s_1^{(m)**} = \frac{4\sqrt{m}}{2n+1} \sum_{l,l'=0}^{r_n} c_l c_{l'} \sum_{s=1}^{n} \sum_{k=1}^{n-s} c_{s,s+k} [e_{s-l} e_{s+k-l'} - \mathbf{E}(e_{s-l} e_{s+k-l'})] \quad (5.64)$$

$$= \frac{4\sqrt{m}}{2n+1} \sum_{l,l'=0}^{r_n} c_l c_{l'} \sum_{h=l-l'+1}^{n-q-l'} \sum_{q=1-l}^{n-l} c_{q+l,q+h+l'} [e_q e_{q+h} - \mathbf{E}(e_q e_{q+h})].$$

We consider the terms associated with $h = 0$. By using the assumption of the fourth-order moments, $\mathbf{E}[(e_q e_{q+h} - \mathbf{E}(e_q e_{q+h}))^2]$ is bounded and $|c_{st}| \leq 2$ $(s, t = 1, \ldots, n$, the number of terms with $h = 1$ is $O(n)$. Hence, the effects of terms associated with $h = 0$ are stochastically negligible as $n \to \infty$.

Since some parts of the above summation (i.e., the terms in $\sum_{l-l' \leq h < 1}[\cdot]$, $\sum_{n-q-l' \leq h < n-q}[\cdot]$, $\sum_{1-l \leq q < 0}[\cdot]$, and $\sum_{n-l \leq q \leq n-1}[\cdot]$) can be of negligible order asymptotically, we can approximate the summation as

$$s_{11}^{(m)***} = \frac{4\sqrt{m}}{2n+1} \left[\sum_{l=0}^{r_n} c_l \sum_{l'=0}^{r_n} c_{l'} \right] \sum_{h=l-l'+1}^{n-q-l', h \neq 0} \sum_{q=1-l}^{n-l} c_{q+l,q+h+l'} e_q e_{q+h} \quad (5.65)$$

$$\sim \frac{4\sqrt{m}}{2n+1} \left[\sum_{l=0}^{r_n} c_l \sum_{l'=0}^{r_n} c_{l'} \right] \sum_{h=1}^{n-q} \sum_{q=1}^{n} c_{q+l,q+h+l'} e_q e_{q+h},$$

where we denote $c_{s,t} = 0$ $(s > n$ or $t > n)$ for the notational convenience.

(**Step 5**): As the final step with $p = 1$, we approximate the sequence of weakly dependent random variables by a sum of independent noise random variables, and apply the CLT.

Let $m_n = [n^\alpha]$ $(0 < \alpha < 0.8)$, $K_n = [n/m_n]$, $N_n = [n^{\delta/2}]$ $(\delta > 0)$, and $M_n = [n^{1-\delta/2}]$ such that $1 - \delta/2 > 0$ and $\alpha + \delta/2 > 1$. Then, $K_n/N_n \to 0$, $N_n/n \to 0$ $\sqrt{m_n}/n \sim [1/\sqrt{n}][1/\sqrt{K_n}]$ and $M_n \sim n/N_n$ as $n \to \infty$. In the following we utilize the relation $c_{q+l,q+h+l'} - c_{q,q+h} = o(1)$ for $l, l' = 1, \ldots, r_n$ if we take r_n such that $r_n \times m_n/n \to 0$ as $n, m_n \to \infty$. This is because

$$\sin 2\pi m \left[\frac{2q+h+l+l'}{2n+1} \right] - \sin 2\pi m \left[\frac{2q+h}{2n+1} \right]$$

$$= \sin 2\pi m \left[\frac{2q+h}{2n+1} \right] \left[\cos 2\pi m \left(\frac{l+l'}{2n+1} \right) - 1 \right]$$

$$+ \cos 2\pi m \left[\frac{2q+h}{2n+1} \right] \sin 2\pi m \left[\frac{l+l'}{2n+1} \right] \to 0$$

as $n \to \infty$.

Furthermore, by using that some parts of (5.65) are of smaller orders as $n \to \infty$ (the terms in $\sum_{h=K_n+1}[\;\cdot\;]$), we can apply the CLT to

$$
s_{11}^{(m)****} = 2 \left[\sum_{l=0}^{r_n} c_l \right]^2 \frac{1}{\sqrt{n}} \frac{1}{\sqrt{K_n}} \sum_{q=1}^{n} \sum_{h=1}^{K_n} c_{q,q+h} e_q e_{q+h}, \tag{5.66}
$$

where we denote $c_{q,q+h} = 0\ (q+h > n)$ for notational convenience. We notice that $c_{q,q+h}\ (q = 1, \ldots, n)$ is a sequence of bounded real numbers.

Let

$$
V_{qn} = \frac{1}{\sqrt{K_n}} \sum_{h=1}^{K_n} c_{q,q+h} e_q e_{q+h} \tag{5.67}
$$

and

$$
U_{jn} = \frac{1}{\sqrt{N_n}} [V_{(j-1)N_n+1,n} + \cdots + V_{jN_n-K_n,n}] \quad (j = 1, \ldots, M_n). \tag{5.68}
$$

Then, we find that $\mathbf{E}[V_{q,n}] = 0$, $\mathbf{E}[V_{q,n} V_{q+h,n}] = 0$ (h is any non-zero integer), $\mathbf{E}[V_{q,n}^2]$ are bounded. Further, we have that $U_{1,n}, \ldots, U_{M_n,n}$ are mutually independent and $\mathbf{E}[U_{i,n}^4]\ (i = 1, M_n)$ are uniformly bounded using the assumption of the boundedness of the fourth-order moments of $V_q\ (q = 1, \ldots, n)$. Since other terms except the leading term are stochastically of the smaller order, we can ignore them for evaluating the limiting distribution, and we apply the Lyapunov-type CLT. By using the relation that

$$
\frac{1}{\sqrt{n}} \sum_{q=1}^{n} V_{qn} - \frac{1}{\sqrt{M_n}} \sum_{j=1}^{M_n} U_{jn} \xrightarrow{p} 0 \tag{5.69}
$$

as $n \to \infty$. The remaining terms in the above approximation are of smaller order (i.e., K_n terms in each $U_{jn}\ (j = 1, \ldots, M_n)$) when $m_n, n \to \infty$ and $m_n = [n^\alpha], 0 < \alpha < 0.8$ because of the condition $K_n/N_n \to 0$. Then we have the asymptotic normality of (5.66) when $p = 1$. By using the relation $m \sum_{s,t=1}^{n} c_{st}^2 = (n + 1/2)^2$ and

$$
4 \left[\sum_{j=-\infty}^{\infty} c_j \right]^2 \sum_{g=1}^{n} \sum_{h=1}^{K_n} c_{g,g+h} [\sigma_v^{(x)}]^4 \sim 2 \left[\sum_{j=-\infty}^{\infty} c_j \right]^2 \sum_{s,t=1}^{n} c_{st}^2 [\sigma_v^{(x)}]^4, \tag{5.70}
$$

we have the desired result of the asymptotic variance when $p = 1$.

(**Step 6**): When $p \geq 1$, we take $\mathbf{a}'\mathbf{C}_s^{(x)} = c_s$, $\mathbf{a}'\mathbf{v}_s^{(x)} = v_s^{(x)}$, and $\mathbf{a}'\mathbf{v}_s = v_s$ for any $p \times 1$ vector $\mathbf{a}$. Then, we evaluate the asymptotic covariance by calculating the covariance of $\sum_{q,h} c_{q,q+h} \boldsymbol{\gamma}_a' \mathbf{v}_q \boldsymbol{\gamma}_b' \mathbf{v}_{q+h}$ and $\sum_{q',h'} c_{q',q'+h'} \boldsymbol{\gamma}_c' \mathbf{v}_{q'} \boldsymbol{\gamma}_d' \mathbf{v}_{q'+h'}$, where $\boldsymbol{\gamma}_a$ represents any constant $p \times 1$ vector. Then, after straightforward

evaluations, we finally find the asymptotic covariance in Theorem 5.3 as $\sigma_{ac}^{(x)}\sigma_{bd}^{(x)} + \sigma_{ad}^{(x)}\sigma_{bc}^{(x)}$ $(a, b, c, d = 1, \ldots, p)$.
(Q.E.D)

Proof of Theorem 5.1: We use the representation

$$\sqrt{m_n}[\hat{\boldsymbol{\beta}}_m - \boldsymbol{\beta}] = \sqrt{m_n}\mathbf{G}_{22}^{*-1}(\mathbf{0}, \mathbf{I}_k)\mathbf{G}_m^*\begin{pmatrix} 1 \\ -\boldsymbol{\beta} \end{pmatrix}, \tag{5.71}$$

where $k = p - 1$.

Because $\mathbf{G}_{22}^* \overset{p}{\to} \boldsymbol{\Sigma}_{22}^{(x)}$ $(m/n \to 0, n \to \infty)$ and under the assumption that $\boldsymbol{\Sigma}_{22}^{(x)}$ is a positive definite matrix, we investigate the asymptotic distribution of

$$\sqrt{m_n}[\hat{\boldsymbol{\beta}}_m^* - \boldsymbol{\beta}] = \boldsymbol{\Sigma}_{22}^{(x)-1}\frac{1}{\sqrt{m}}(\mathbf{0}, \mathbf{I}_k)\mathbf{G}_m^*\begin{pmatrix} 1 \\ -\boldsymbol{\beta} \end{pmatrix}, \tag{5.72}$$

which is asymptotically equivalent to (5.71). Then, its asymptotic variance-covariance matrix can be written as

$$\mathbf{AV}[\hat{\boldsymbol{\beta}}_m] = \boldsymbol{\Sigma}_{22}^{(x)-1}\mathbf{Cov}\left[(\mathbf{0}, \mathbf{I}_k)\mathbf{Sb}, \ \mathbf{b'S}\begin{pmatrix} \mathbf{0'} \\ \mathbf{I}_k \end{pmatrix}\right]\boldsymbol{\Sigma}_{22}^{(x)-1}, \tag{5.73}$$

where $\mathbf{S} = \sqrt{m_n}[\mathbf{G}_m^* - \boldsymbol{\Sigma}_x]$ $(= (s_{jk}))$ and $\mathbf{b} = \begin{pmatrix} 1 \\ -\boldsymbol{\beta} \end{pmatrix}$ $(= (b_j))$.

By using Theorem 5.3, we can evaluate the (l, l')-th element $(l, l' = 2, \ldots, k+1 = p)$ of $\boldsymbol{\Sigma}_x = (\sigma_{l,l'}^{(x)})$ as

$$\mathbf{Cov}\left[\sum_{j=1}^{k+1} b_j s_{jl} \sum_{j'=1}^{k+1} b_{j'} s_{j'l'}\right] = \sum_{j,j'=1}^{k+1} b_j b_{j'}\left(\sigma_{j,j'}^{(x)}\sigma_{l,l'}^{(x)} + \sigma_{j,l'}^{(x)}\sigma_{j',l}^{(x)}\right)$$

$$= \sigma_{l,l'}^{(x)}\sum_{j=1}^{k+1} b_j\left[\sum_{l'=1}^{k+1} b_{j'}\sigma_{j,j'}^{(x)}\right] + \left[\sum_{j=1}^{k+1} b_j\sigma_{j,l'}^{(x)}\right]\left[\sum_{j'=1}^{k+1} b_{j'}\sigma_{l,j'}^{(x)}\right]$$

$$= \sigma_{l,l'}^{(x)}\sigma_{11.2}^{(x)}$$

because $[\sigma_{21}^{(x)}, \boldsymbol{\Sigma}_{22}^{(x)}]\mathbf{b} = \mathbf{0}$ and

$$\left[\sigma_{11}^{(x)}, \sigma_{12}^{(x)}\right]\mathbf{b} = \sigma_{11}^{(x)} - \sigma_{12}^{(x)}\boldsymbol{\Sigma}_{22}^{(x)-1}\sigma_{21}^{(x)}. \tag{5.74}$$

Then we have the result of the asymptotic variance-covariance matrix of (5.71) in Theorem 5.1.
(Q.E.D)

Proof of Theorem 5.2 We use the representation

$$\hat{\mathbf{B}}_m - \mathbf{B} = (\mathbf{W}_m^{*'}\mathbf{W}_m^*)^{-1}\mathbf{W}_m^{*'}\mathbf{U}_m^*, \qquad (5.75)$$

where $\mathbf{U}_m^* = \mathbf{J}_m\mathbf{P}_n\mathbf{C}_n^{-1}\mathbf{U}_n$. Rewrite (5.19) as $\sqrt{m_n}[\hat{\mathbf{B}}_m - \mathbf{B}] = (\frac{1}{m_n}\mathbf{W}_m^{*'}\mathbf{W}_m^*)^{-1}\frac{1}{\sqrt{m_n}}\mathbf{W}_m^{*'}\mathbf{U}_m^*$. By using a similar argument as the proof of Theorem 5.1 under the assumption of (5.1) and (5.2), and $\mathbf{w}_i$ are instrumental variables (and exogenous), we find that

$$AV[\hat{\mathbf{B}}_m] = \mathbf{\Sigma}_{w^*}^{-1}\mathbf{Cov}\left[\frac{1}{\sqrt{m}}\mathbf{W}_m^{*'}\mathbf{U}_m^*, \frac{1}{\sqrt{m}}\mathbf{W}_m^{*'}\mathbf{U}_m^*\right]\mathbf{\Sigma}_{w^*}^{-1}. \qquad (5.76)$$

Then, by using Theorems 5.3 and 5.1 we have the result.
(**Q.E.D**)

References

Anderson TW (1971) The statistical analysis of time series. Wiley
Anderson TW (1984) Estimating linear statistical relationships. Ann Stat 12:1–45
Anderson TW (2003) An introduction to multivariate statistical analysis, 3rd edn. Wiley
Baxter M, King RG (1999) Approximate band-pass filters for economic time series. Rev Econ Stat 81–4:575–593
Brillinger D, Hatanaka M (1969) An harmonic analysis of nonstationary multivariate economic processes. Econometrica 35:131–141
Brockwell P, Davis R (1990) Time series: theory and methods, 2nd edn. Wiley
Census Bureau (2020) X-13ARIMA-SEATS reference manual (Accessible HTML Outout Version), U.S. Census Bureau. https://www.census.gov/data/software/x13as.html
Engle R (1974) Band spectrum regression. Econometrica 15–1:1–11
Fuller W (1987) Measurement error models. Wiley
Granger C, Hatanaka M (1964) Spectral analysis of economic time series. Princeton University Press
Hosoya Y (1997) A limit theory for long range dependence and statistical inference on related models. Ann Stat 25(1):105–137
Kunitomo N, Sato S (2021) A robust filtering method for noisy nonstationary multivariate time series with econometric applications. Open Access Japanese J Stat Data Sci 4:373–410, Springer
Müller U, Watson M (2018) Long-run covariability. Econometrica 86–3:775–804

Chapter 6
A Filtering Method of Constructing Macro Consumption Index

Abstract We consider practical issues when we use monthly, quarterly, and annual macroeconomic time series that are published regularly with different frequencies by several government agencies in Japan. In order to address the issue with the multivariate economic time series, we apply the SIML filtering based on the range of time series that are non-stationary with seasonality and noise. We apply the SIML filtering method to construct macro consumption indicators using quarterly and monthly macro consumption series.

6.1 Introduction

This chapter focuses on recent Japanese quarterly and monthly consumption series as a typical application of the SIML filtering. We applied the SIML filtering method to handle macro consumption time series with different frequencies, which are regularly published as official statistics by different central agencies in Japan. Regarding Japanese macro consumption, the GDP-final-consumption series is published by ESRI, the Cabinet Office of Japan only once each quarter.[1] The monthly household budget survey is published by the Statistics Bureau of Japan, while commercial statistics and tertiary industry statistics are published by the Ministry of Economy, Trade and Industry (METI). Additionally, the consumption trend index (CTI) has been constructed and published as a monthly index series by the Statistics Bureau of Japan. Therefore, to understand the movement of macro consumption in Japan, it is important to consider the various official time series comprehensively. The objective of this chapter is to propose a statistical method to estimate the true consumption state variable from non-stationary economic time series. We will illustrate the practical application of the SIML filtering method by demonstrating its use in constructing a monthly macro consumption index from monthly and quarterly consumption data with different frequencies.

---

[1] A revision of quarterly (quick) GDP-estimate is reported regularly after a month. It is mainly because new fixed-investment data become available from Policy Research Institute, the Ministry of Finance.

6.2 An Empirical Problem

In order to gain insight into recent macroeconomic consumption movement in Japan, two data sources are usually used: data on consumption trends in household budget surveys based on sample survey of households and commercial statistics based on sample survey of private companies. However, there are considerable discrepancies between consumption trends in spending and sales obtained from *Kakei-Chosa* (household budget surveys), *Shogyo-Doutai-Toukei* (commercial statistics), and *Dai-Sanji-Sangyo-Toukei* (consumption index of tertiary industry). Understanding the main cause of this discrepancy and identifying a correct way to construct macro consumption indices is an important issue in economic statistics. On the one hand, household consumption data based on sample surveys of households are constructed as an aggregate value. When interpreting the household budget survey data, recent household trends in Japan, such as changes in family size and number of households, must be considered because the typical household size has been gradually diminishing. The household budget survey is a monthly survey that includes detailed items on expenditures and financial activities, and the sample size is approximately 8,000 to 9,000 monthly based on the three-stage random sampling with rotation. On the other hand, consumption data from commercial statistics are obtained from companies, through survey data on the supply side of production and sales, using a random sampling approximately with 20,000 firms monthly. Because the survey is conducted on a business establishment basis, consumption components, such as corporate consumption, government consumption, and inbound consumption, are included. This information is not necessarily consistent with the figures related to household consumption. Due to these differences, macro consumption measurements such as GDP comprise two types of consumption data: demand-side and supply-side consumption. Then, it is important to understand the characteristics of different consumption concepts and maintain consistency for macro consumption understanding.

Many private and government economists in Japan have emphasized that the most important factor for the macroeconomy is macroeconomic trends; thus, there is an interest in using quarterly GDP-final-consumption figures. According to quarterly GDP estimates constructed at ESRI, the Cabinet office of Japan, final consumption integrates information from both household and corporate consumption and the figures are compiled quarterly. Since economists are interested in monthly fluctuation of the economy, the problem of how to handle monthly and quarterly time series coherently has become an issue. In these data, seasonal adjustment processing is usually performed, meaning most economists use seasonally adjusted data for their judgment. Although they tend to ignore the seasonal time series adjustments, it is quite important to pay attention to them.[2] In addition to quarterly quick estimates of GDP, the Cabinet Office publishes final estimates of GDP annually after considerable time lags. They then revise their seasonally adjusted data because they use more

---

[2] For details of quarterly GDP and its seasonal adjustment procedure, ESRI, Cabinet Office of Japan, National Income Department (2010). The current web-site: https://www.esri.cao.go.jp/en/sna/menu.html.

detailed estimates on the supply side and construct a quarterly series by dividing annual estimates into quarters of some factors such as the government consumption. (The final-GDP figures are constructed by using the benchmark method and other practical techniques.)

The ESRI of the Cabinet Office uses household budget surveys, commercial statistics, and the tertiary industry activity index to construct preliminary figures for quarterly GDP consumption. Monthly surveys provide the latest information. For the recent trends in each original series, we show GDP-final-consumption (Fig. 6.1), the data from the household survey (the solid lines in Fig. 6.3), the commercial statistics (Fig. 6.5), and tertiary Industry (Fig. 6.7). The fluctuations in these monthly data are different from household consumption; thus, it may be difficult to fully understand macro consumption behavior using these sample survey data. For example, consumption data on the sales side show new forms of consumption, such as inbound consumption. There may be some problems with the survey data's accuracy that should not be ignored. While we can utilize monthly data aggregated at the monthly level, trends of GDP consumption are estimated at the quarterly and annual levels. Thus, we need to understand consistently the data reported and grasp Japanese macro consumption trends.

Using statistical time series analysis, it may be possible to integrate monthly data and the information obtained at the quarterly level without contradiction. In this chapter, we discuss one construction method of macro-index from multidimensional time series when some observations are missing and the data are incomplete. GDP consumption is calculated as an aggregate value of various data on a quarterly basis and we regard monthly GDP data as a time series with missing values. Because of monthly data on consumption available, we consider the case when a part of the multidimensional time series data is missing and we try to estimate the missing values as the state variables.

Although economists and policy makers are usually interested in the trend-cycle components of economic time series, actual economic time series contain not only the trend-cycle components, but also seasonal components, irregular components, outliers, and changing points in many cases. Because several underlying factors are observed in one time series, the moving average method and/or the regression method was often traditionally used. The official statistics agencies of many countries including Japan use the seasonal adjustment method X-12-ARIMA or X-13ARIMA-SEATS, which is based on moving average procedure. Then, we should remember that X-12-ARIMA and X-13ARIMA-SEATS use a univariate time series regression model called the Reg-ARIMA model in the internal calculation. For each published series in official statistics, the practice of official agencies in Japan is that each official statistician in charge of a series independently adopts the Reg-ARIMA models. In the case of macroeconomic time series, however, there are various relationships between time series, and time series component decomposition is necessary because the different components are mixed, and seasonal adjustments are not necessarily performed in consistent manner across multiple time series. In this chapter, we provide one way to solve these problems. It appears that the SIML filtering method gives robust state estimation results and it can be used for practical applications.

6.3 The Empirical Model

We regard an incomplete time series with missing values as the realization of an observable non-stationary multidimensional time series. The main purpose is to estimate the state of the macro monthly indicator, which is consistent with the quarterly GDP. The consumption series of our information comprises time series data from demand-side household surveys and from supply-side production statistics, which are available as monthly data. The objective variable in the analysis is the GDP-final-consumption series, whose quarterly data are only observed. The quarterly final-consumption data are constructed by combining the monthly demand- and supply-side information. Since the quarterly data are estimated by aggregating the monthly time series on both sides, it is reasonable to assume that there exists a hidden true monthly series.

We have the state estimation problem of the macro consumption index by applying non-stationary multidimensional time series representation. The target (observed) quarterly data is given as y_{1t} ($t = 12(l - 1) + 3j; l = 1, \ldots, n; j = 1, 2, 3, 4; T = 12n$). The observations of monthly data are more frequent than quarterly data and we represent a $p - 1$ dimensional vector $\mathbf{y}_{2t}$ ($t = 12(l - 1) + j, l = 1, \ldots, n; j = 1, \ldots, 12; T = 12n$). Then, the statistical problem is that given the available monthly data information we estimate an unobservable state of the true trend/cyclical component x_{1t} ($t = 12(l - 1) + j, l = 1, \ldots, n; j = 1, \ldots, 12$). Here, in this chapter we use T as the sample size and n as the number of years of data because we have 4 or 12 observations in each year.

For non-stationary multidimensional time series with noise when the observations of some variables are incomplete, the estimation algorithm for optimal state estimation is not known. We propose a new method that can be realized easily and is robust using trend-cyclical factors, irregular fluctuation factors, and seasonal fluctuation factors. We denote the total number of variables $p = (p_1 + p_2)$ as the number of target variables p_1 in the general case. Although it is possible to generalize as the number of explanatory variables p_2, in this analysis we use the case of $p = 4$, $p_1 = 1$, and $p_2 = 3$.

First, from the available quarterly data we estimate the relationship between trend-cyclic (TC) variables. For this purpose, it is necessary to first remove seasonal and noise components from the original data before estimating the TC components.

(i) As a TC component, we extract periodic components over 1 year using SIML filtering. Specifically, let m be much larger than the seasonal cycle and apply SIML filtering. By this operation it is possible to remove seasonal fluctuations and irregular components when only quarterly data are available. As the first round estimates, we divide the values by 3 as the monthly series from the corresponding quarterly series and apply SIML filtering. This simple method allows removing some seasonal and irregular components from the monthly series.

(ii) Next, for the observed quarterly data $\mathbf{y}_{1t}$ ($t = 12(l - 1) + 3j, l = 1, \ldots, n; j = 1, 2, 3, 4$) we set the estimate of state variable be $\hat{\mathbf{x}}_{1t}$ ($t = 12(l - 1) +$

$3j, l = 1, \ldots, n; j = 1, 2, 3, 4).$ (Set the objective state variable to $\mathbf{x}_{1t}$ $(t = 12(l - 1) + 3j, l = 1, \ldots, n; j = 1, 2, 3, 4).)$ Because the series for which monthly data are available $y_{2,t}, y_{3t}, y_{4t}$, we set the estimate of the state variables x_{2t}, x_{3t}, x_{4t} $(t = 12(l - 1) + j, l = 1, \ldots, n; j = 1, \ldots, 12).$

We then estimate the relationship between the TC components of the monthly series. We note that $\mathbf{x}_t$ is a non-stationary state variable and the estimate of variable x_{1t} $(t = 12(l - 1) + 3j, l = 1, \ldots, n; j = 1, 2, 3, 4)$ is a three-dimensional explanatory variable vector $\mathbf{x}_t^{(d)} = (x_{2t}, x_{3t}, x_{4t})'$ $(t = 12(l - 1) + 3j, l = 1, \ldots, n; j = 1, 2, 3, 4).$ We run regression, and let the regression coefficient vector be $(b_2, b_3, b_4)'$. The relationship between the state variable vectors is represented by the least squares coefficients of the regression method, which are written as the vector $\mathbf{b} = (b_2, b_3, b_4)'$. We write the true coefficient as $\boldsymbol{\beta} = (\beta_2, \beta_3, \beta_4)' = \boldsymbol{\beta}'_x$ and

$$x_{1t} = \beta_2 x_{2t} + \beta_3 x_{3t} + \beta_4 x_{4t} + u_t. \tag{6.1}$$

The objective variable is the monthly indicator and it is a regression model without a constant term. The coefficient vector can be interpreted as a co-integration vector and the error term u_t can be treated as a stationary series with the initial value as zero.

Alternatively, we could use regression with differenced data. In this case, the estimated value of the difference variable Δx_{1t} $(t = 12(l - 1) + 3j, l = 1, \ldots, n; j = 1, 2, 3, 4)$ is a three-dimensional vector explanatory variable vector $\Delta \mathbf{x}_t^{(d)} = (\Delta x_{2t}, \Delta x_{3t}, \Delta x_{4t})'$ $(t = 12(l - 1) + 3j, l = 1, \ldots, n; j = 1, 2, 3, 4).$ The regression coefficient vector is $(b_2^*, b_3^*, b_4^*)'$. A statistical model that explains trends in terms of levels and differences can also be considered as we have reasonable estimates using the original data without the differencing transformation in our analysis.

(iii) Third, we consider the effects of the outliers and changing points using dummy variables. It is possible to estimate the target state variable by using the estimated coefficients. The error evaluation is performed by estimating the state variable at every time when new data are obtained.

$$\hat{x}_{1t} = b_2 x_{2t} + b_3 x_{3t} + b_4 x_{4t} \quad (t = 12(l - 1) + j; l = 1, \ldots, n; j = 1, \ldots, 12) \tag{6.2}$$

By repeating this procedure, the state estimate of the objective variable can be obtained.

By estimating the state variables using the SIML filter, we find the trend-cyclical component, seasonal component, and noise component. Thus, it becomes possible to estimate the state variables, although there is some difficulty to estimate the state variables of the monthly seasonal component. To remove the monthly noise component, it may be preferable to use the estimated TC component. It is more robust and may be suitable as the monthly consumption indicator. Although the regression coefficients are stable we use the level variables, it may be difficult to imagine that they are constant over time. Every time the quarterly data become available, they

Fig. 6.1 GDP-final-consumption (we constructed monthly series by interpolation and we set $m = 20$)

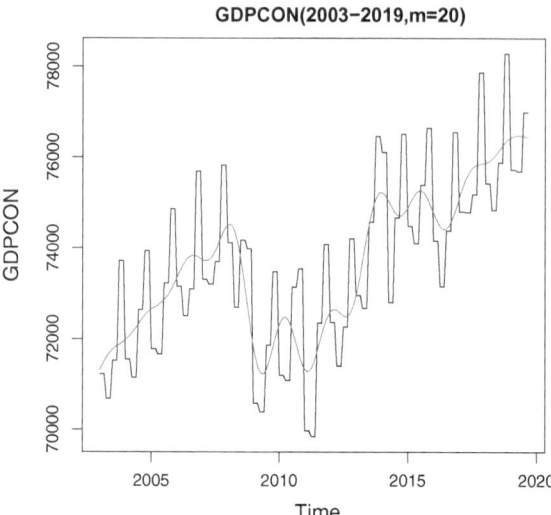

should be updated. By repeating the statistical filtering and smoothing, it may be possible to construct macro consumption indicators that are coherent in whole.

We have used a statistical regression model for analyzing the relationship between time series, but the data are apparently non-stationary. Hence, we should be cautious about the estimated results because the justification for the estimation method for ordinary regression models is not applicable. We also note that because it is a factor decomposition of a non-stationary series when using a difference series, we have avoided the possibility that the MA unit root may be mixed with the error term. Although a certain degree of fit can be obtained, the estimation results are not necessarily reliable.[3]

6.4   The Empirical Result

Figure 6.1 shows the quarterly data on GDP real final expenditure, which is constructed as a monthly time series. A smoothed curve is constructed by estimating the state of the trend-cyclical part with $m = 20$ by the SIML filtering. Since monthly data are not actually available, they are calculated as if monthly consumption was available. We need an initial estimate of the monthly trend-cyclical component. When monthly data are constructed from quarterly data, the monthly trend-cyclical component fluctuates smoothly. Spectral information from this monthly series is shown in Fig. 6.2.

---

[3] For further discussion here, see Chap. 10 of Hayashi (2000).

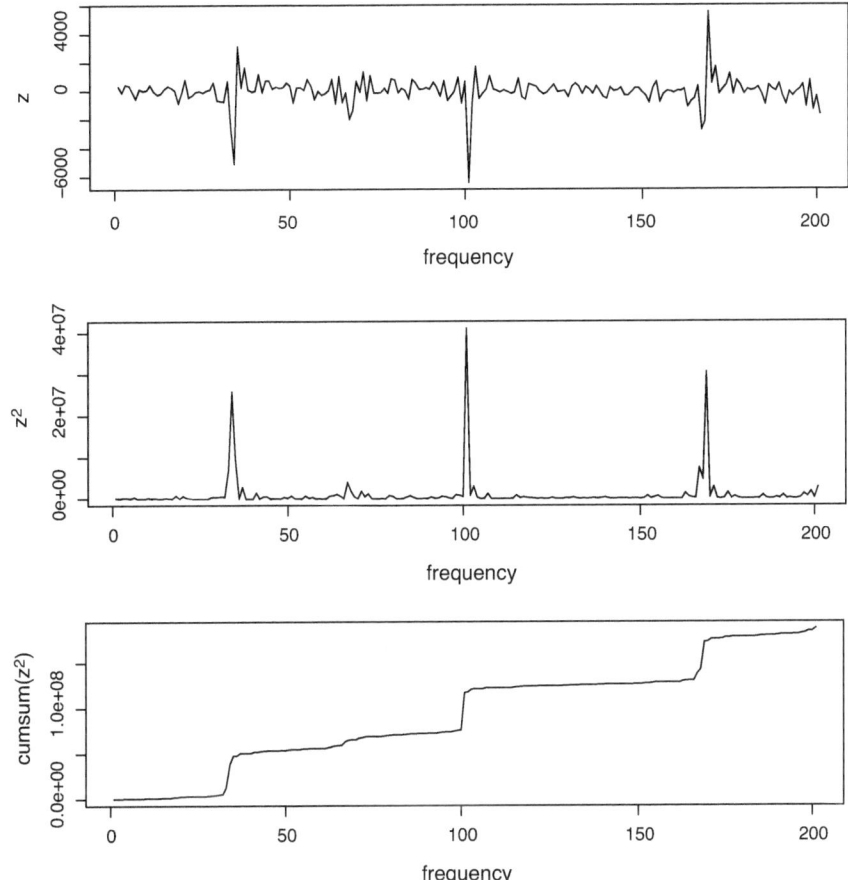

Fig. 6.2 Spectral decomposition of GDP-final-consumption (original data are the quarterly real GDP and monthly real GDP are constructed by interpolation)

Next, we present several auxiliary series, monthly series of household consumption as shown in Figs. 6.3 and 6.4, the monthly series of commercial statistics as shown in Figs. 6.5 and 6.6, and the monthly series of the tertiary industry index as shown in Fig. 6.7. In each figure, the estimates of the trend-cycle components are shown as smoothed curves. Using the SIML transformation, we obtain the spectral information from the monthly series of household expenditures in Fig. 6.4, the spectral information from the monthly series of commercial dynamics statistics in Fig. 6.6, and the spectral information from the monthly series of the tertiary index (Fig. 6.8).

We state some observations from these series and figures.

(i) On the household budget survey the levels of household consumption and trend components are slightly decreasing. The level of the commercial dynamics

Fig. 6.3 Figure 6.2:
Kakei-Shohi (monthly
family expenditure series
published by Statistics
Bureau)

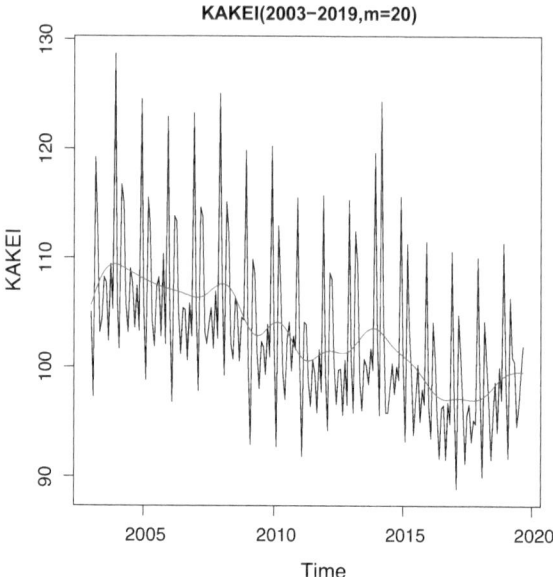

series and tertiary industry and their trends are slightly increasing. These states
of trend components and cyclical components are estimated via the SIML
filtering.

(ii) Although the cyclical components of monthly series are related, time series
for each source of data fluctuates differently (For example, seasonal fluctuation components associated with the sampling scheme and irregular fluctuation
components). The quarterly GDP-final-consumption series is understood as a
composite of the cyclical parts of the monthly three series.

(iii) We have chosen $m = 20$ for the state estimation of trend-cyclical components
for each series by SIML filtering. We do not observe that there is much influence
of seasonal or irregular components by this choice.

From using the monthly three series, we obtain reasonable trend-cyclical fluctuations by using the SIML filtering.

We have used data from January 2003 to December 2018 and we report the
empirical results when we set $m = 36$ for the original series. The estimation results
of the regression analysis on the level variables are summarized in Table 6.1. The
result may be appropriate to maintain consistency with the estimated states. The fit
of the regression is good, but it may be due to the non-stationarity of consumption
time series. Because the level time series is used, the t-value does not follow the
t-distribution under the standard hypothesis. It should be understood as information
that has a meaning indicating its significance. We should also note that the monthly
series obtained from quarterly GDP data was used as the dependent variable and it
differs from the standard regression analysis.

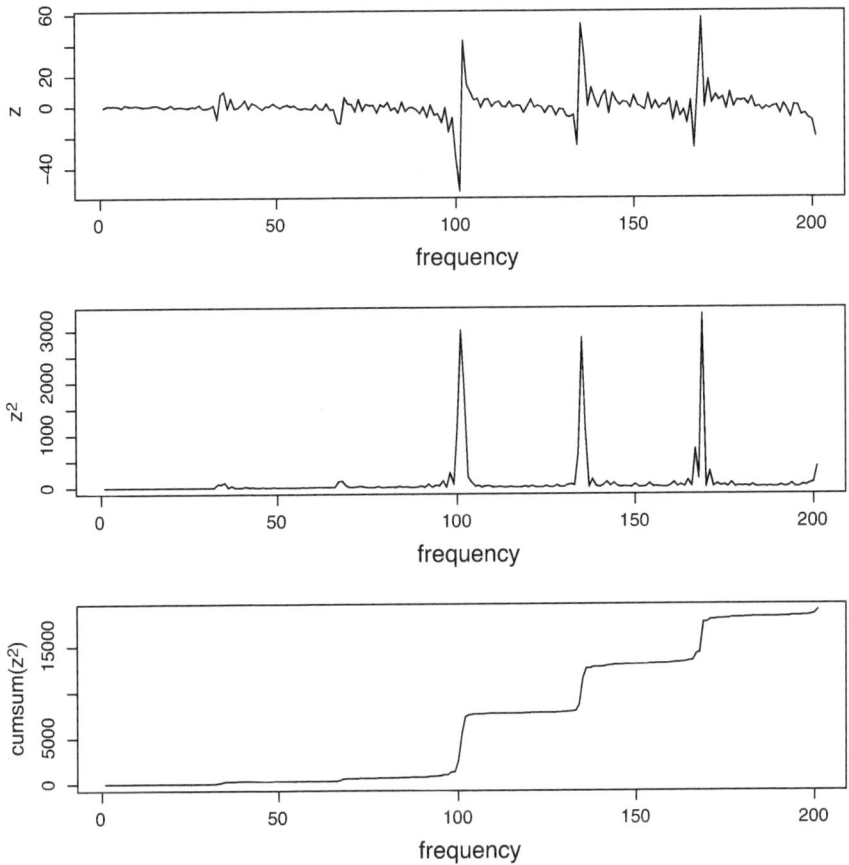

Fig. 6.4 Kakei-Shohi ($\{z_k\}$ series, $\{z_k^2\}$ series, and the cumulative $\{z_k^2\}$ series)

Next, we examine trends and cyclical components of the data from 2002 to 2018. We have performed the state estimation and given the coefficients estimated by the regression analysis. Then, we construct the consumption index up to 2019 to make comparisons with the values of the CTI of the Statistics Bureau. We have constructed a consumption index series from regression using a difference series given the initial values. January 2017 is set as the reference point, with estimation results shown in Figs. 6.9 and 6.10.

When we examine the estimated series of (TC) + (noise series), we have a noisy monthly series, even though the quarterly estimates are quite close to the published values. This may be due to the different seasonality and irregular fluctuations observed in each time series. We also find that the estimated trend-cyclical components are close to the CTI published time series. In each series there are some abnormal values and changing points and monthly fluctuations reflecting irregular fluctuations. Then, in some cases, there is considerable deviation from the CTI pub-

Fig. 6.5 Commercial
statistics (monthly
commercial consumption
series published by METI)

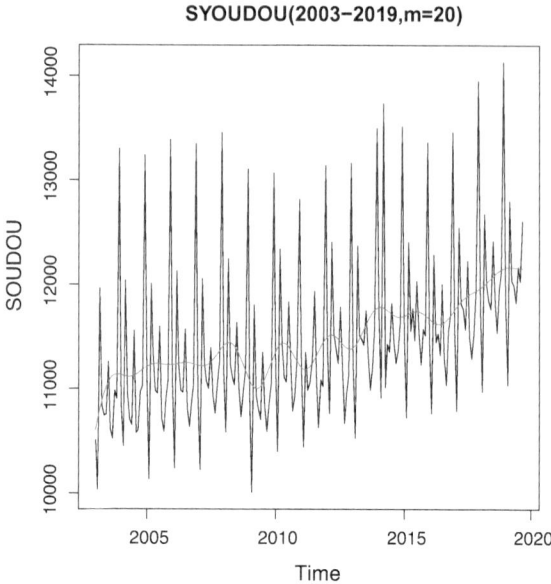

Table 6.1 Regression result ($m = 36$)

	(Kakei-Chosa)	(Commercial statistics)	(Tertiary industry index)
(Coefficient)	138.601	4.562	53.562
(SD)	7.379	0.384	40.343
(t-value)	18.783	11.887	1.322

lished index. There are several reasons for this phenomenon including the stability
of regression coefficients and handling of abnormal values and changing points. It
seems that there are change points suggesting the need for periodic updates of the
regression coefficients. (For the detailed discussion, see Kunitomo et al. (2022).) By
comparing the estimated consumption series with the CTI published consumption
index, we have several observations. For instance, the event occurred in March 2011,
while the Great East Japan Earthquake and the consumption tax hike took place
in April 2014. The effects of these events on consumption are noticeable. In the
CTI published series, these large fluctuations may impact the estimation of trend-
cyclical components. The temporary effects and the effects of major changes can
be also treated by SIML filtering. Using appropriate dummy variables, we perform
frequency regression by incorporating the outliers and change points to estimate the
monthly index. In our analysis, we used a simple AO type consumption tax dummy
variable and we found that the effect is very large. Thus, it may be necessary to
further consider the use of variables such as the Ramp-dummy variable.

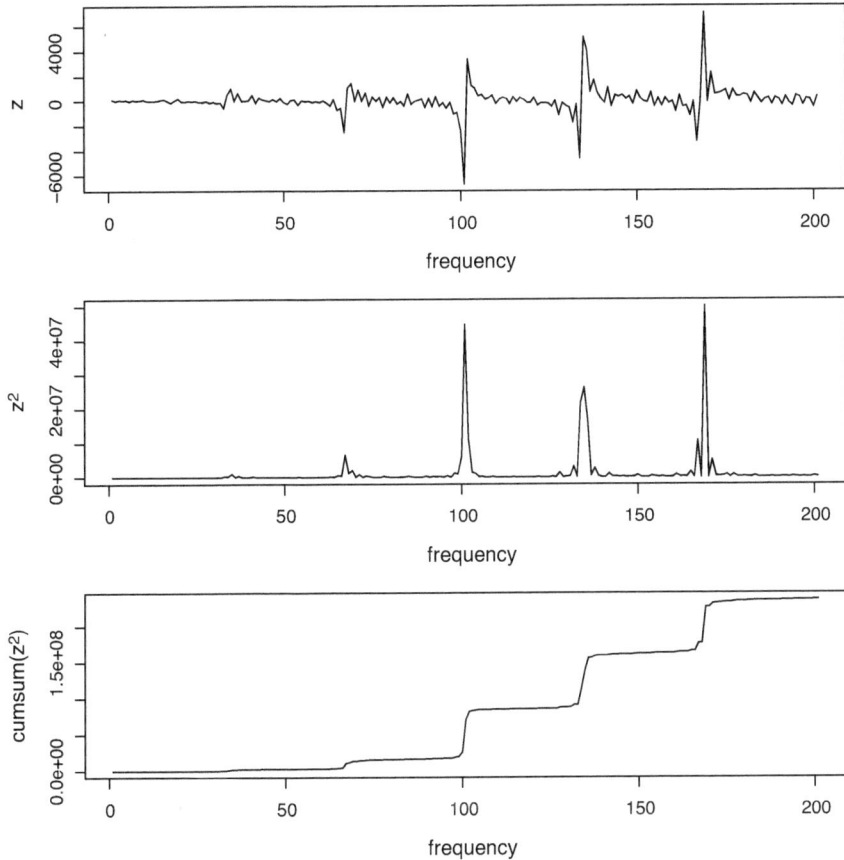

Fig. 6.6 Commercial statistics, spectral decomposition ($\{z_k\}$ series, $\{z_k^2\}$ series, and the cumulative $\{z_k^2\}$ series)

Finally, using dummy variables for the time series, the monthly consumption series for the CTI state of the monthly TC series is shown in Fig. 6.11. This consumption series is based on the TC series using regression with $m = 36$ (monthly) with the dummy variables for the Lehman-Shock, the Great East Japan Earthquake, and the introduction of consumption tax with a type of Ramp-dummy variable.[4] It is necessary to consider various possibilities, including the day-of-the-week effect and we used the model selection and filter selection. In that sense, there are still many practical issues. Our overall evaluation of the empirical analysis is that the quarterly and monthly series constructed using SIML provides a satisfactory performance for practical use.

---

[4] The dummy variable *tkramp*, one type of double ramp variable (Example 5.4), was used by Makoto Takaoka (see Takaoka (2015)).

Fig. 6.7 Indices of tertiary industry activity (monthly tertiary consumption series published by METI)

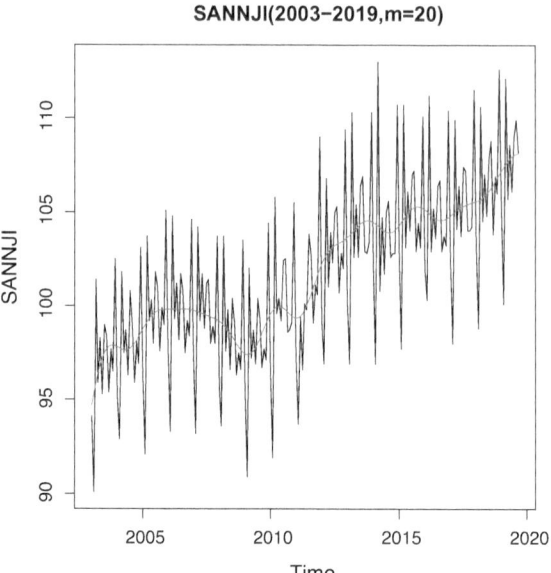

According to the manual of the Statistics Bureau, calculating the current official CTI including the seasonal adjustments is quite complicated. Hence, because our approach is fairly simple and results in a similar time series, there is potential for improvement over the current CTI.

6.5 Some Remarks

This chapter considers the practical situation of how to deal with non-stationary time series containing noise with yearly, quarterly, or monthly frequencies. There are some difficulties in achieving consistency with each other when we observed time series data at different frequencies. By considering the state variable estimation and smoothing problem, we confirmed that the SIML filtering is promising. As a real example, we consider the situation where our target is to construct a monthly macro consumption indicator and the target time series is the quarterly consumption index while monthly time series data are available as incomplete observations. The method proposed here is quite simple and we can construct a time series index which is similar to the published series of monthly TCI series. According to our method, we can treat trend-cycle components observed in many macro time series and treat seasonal fluctuation factors, irregular fluctuation factors, outliers, and change-point factors. These factors are related to non-stationary time series data, which can be easily incorporated by using the SIML filtering, and the meaning of the seasonal adjustment series is expected to become clearer.

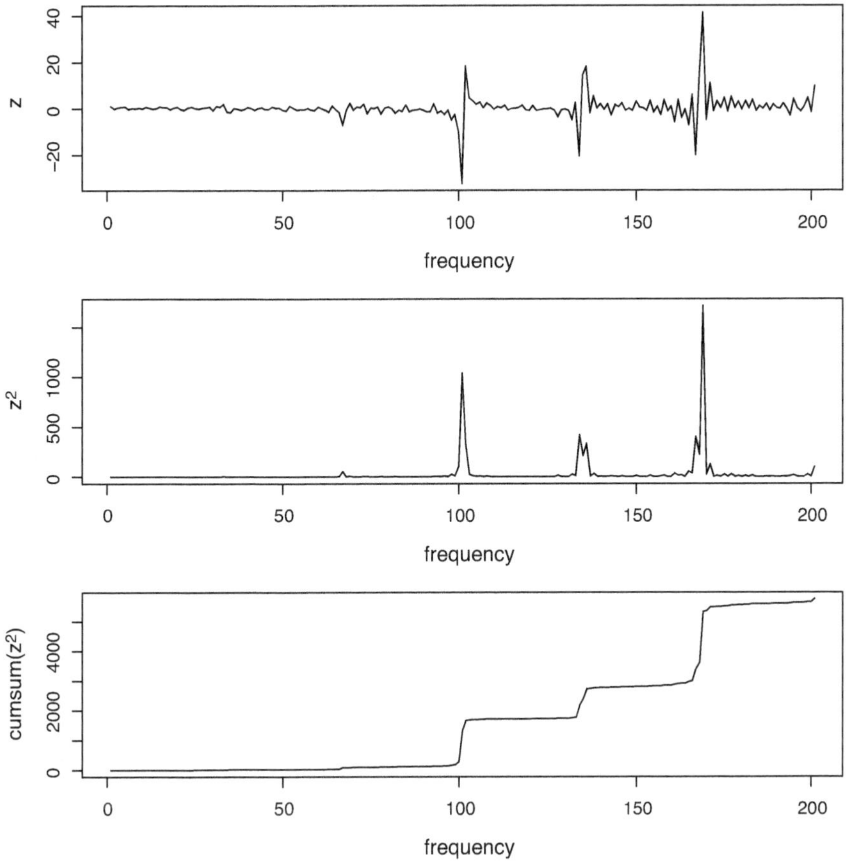

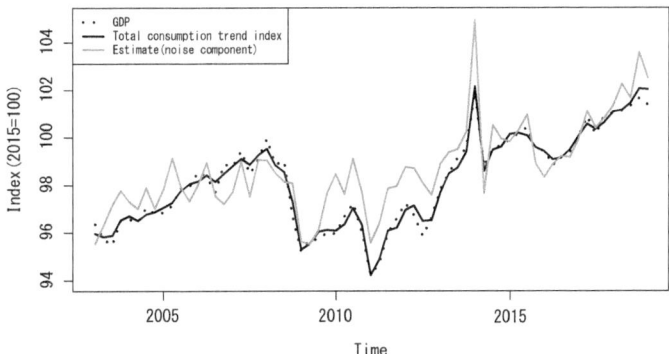

Fig. 6.8 Indices of tertiary industry activity, and spectral decomposition ($\{z_k\}$ series, $\{z_k^2\}$ series, and the cumulative $\{z_k^2\}$ series)

Fig. 6.9 Consumption index (the estimated consumption index series and estimated trend-cycle series)

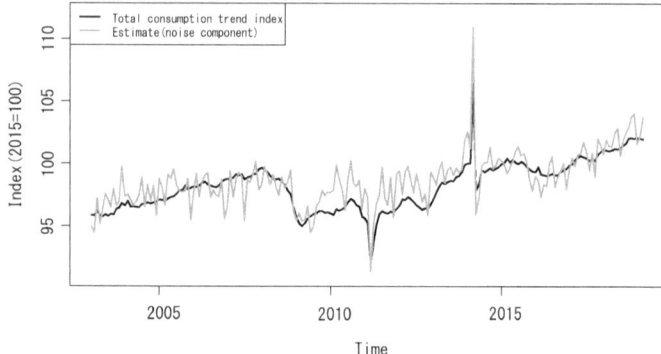

Fig. 6.10 Consumption index (the estimated consumption index series and estimated trend-cycle series)

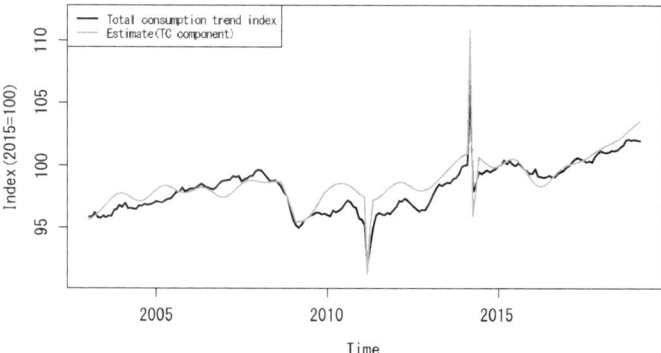

Fig. 6.11 Consumption index (the estimated consumption index series and estimated trend-cycle series)

References

Hayashi F (2000) Econometrics. Princeton University Press

Kunitomo, Sakurai, Sato (2022) A filtering method of economic time series and a macro-index (in Japanese). Toukei-Kenkyu-Ihou vol 79. Bureau of Statistics, Japan, pp 1–20

Takaoka M (2015) Economic Time series and seasonal adjustment methods. Asakura-Shoten (in Japanese)

Chapter 7
Conclusions

Abstract In conclusion, the SIML filtering method, which is presented in this publication, exhibits favorable finite sample and asymptotic properties. The simplicity of the SIML filtering renders it a useful tool for practical applications. Furthermore, we highlight additional research avenues that warrant investigation.

In a considerable number of original macroeconomic time series, it is typical to discern the coexistence of non-stationary trends, cycles, seasonal variations, and measurement errors. In addition to the aforementioned components, abrupt changes, trading-day effects, and other irregular components are also observed. It is therefore challenging to remove the seasonal component from the original time series in the seasonal adjustment and construct the macro-index, which involves multiple non-stationary time series. An earlier study on the measurement errors in economic time series was conducted by Morgenstern (1950). For further insight into the historical development of statistical analysis of economic time series, see Nerlove et al. (1995).

In this book, we have developed a new statistical approach to address the issue of non-stationary economic time series. Our methodology employs a frequency domain analysis and the frequency regression based on SIML modeling. We use the SIML method because it also separates the likelihood information of time series data into different frequency parts of their components. Our method sheds a new light on the practical handling method of economic time series, which has been often used in official seasonal adjustments without formal justifications. There are numerous potential applications as evidenced by empirical examples in this book.

There are also further problems to be investigated. This study is based on time series decomposition in Chaps. 2, 3, and 5. There may be more complicated decomposition models including different trend, cycle, and seasonal components. We should investigate the relationships among trends, cycles, and seasonal and irregular noise components of non-stationary and stationary time series in the time and frequency domains. Further extensions of the theorems presented in this book could be developed.

Another issue that requires attention is the computation of the procedure we explained in this book. Most computations reported in this book were executed by

N. Kunitomo and S. Sato, *The SIML Filtering Method for Noisy Non-stationary Economic Time Series*, JSS Research Series in Statistics, https://doi.org/10.1007/978-981-96-0882-9_7

the R-program (called x12simldoc92) developed by Seisho Sato.[1] It is our hope that the texts of this book and the R-program will prove useful in real-world applications.

References

Morgenstern O (1950) On the accuracy of economic observations (1966, revised). Princeton University Press

Nerlove M, Grether DM, Carvalho JL (1995) Analysis of economic time series: a synthesis, Revised edn. Academic Press

---

[1] Please refer to the following link for the latest version:
http://www.kunitomo-lab.sakura.ne.jp/x12simldoc92(kuni2023-2-2).pdf.

Index

© The Author(s), under exclusive license to Springer Nature Singapore Pte Ltd. 2025   117
N. Kunitomo and S. Sato, *The SIML Filtering Method for Noisy Non-stationary
Economic Time Series*, JSS Research Series in Statistics,
https://doi.org/10.1007/978-981-96-0882-9